Environmental Education

Edited & Compiled By

Dr. PREM SUNDER

PRINCIPAL

DARSH COLLEGE OF EDUCATION

GOHANA (SONEPAT), HARYANA

KSK PUBLISHERS & DISTRIBUTORS

7/26, Basement, Ansari Road, Darya Ganj,

New Delhi-110002.

Published by:

KSK PUBLISHERS AND DISTRIBUTORS

7/26, BASEMENT, Ansari Road, Darya Ganj, New Delhi-110002.

Ph.: (O) 011-65190343, 43551324, 42564726
(M) 9811088729, (Fax) 42564726, (Res.) 47091605

E-mail: *khelsahitya1@rediffmail.com*

Web: *www.khelsahitya.trade-india.com*

I.S.B.N. - 978-81-89983-21-5

PRINTED IN INDIA 2009

Laser Typeset by:

JAIN MEDIA GRAPHICS
Delhi-110035.

Printed by:

Vishal Kaushik Printers
A-49, Gali No. 6, Jagatpuri Vistar, Delhi-93

Price: Rs. 600/-

Contents

1

HUMAN POPULATION AND ENVIRONMENT

Population Growth

Population is the number of individual organisms of same species in a given area. Population dynamics refers to the trends of growth of population with time. The study of human population is called demography. Some of the topics studied under demography area the rate and trends of growth of population with time, ratio of females to males, percentages of population in each age group etc.

Exponential Growth and Doubling Time

Exponential growth means growth in the population of organism at a constant rate per unit of time. It is called exponential growth because existing population is multiplied is geometrical ratio.

The geometric pattern of increase is 2, 4, 8, 16 and so on. By contrast, a pattern of growth that increases at a constant amount per unit of time is called arithmetic growth. The sequence is this case might be 1, 2, 3, 4, or 1, 3, 5, 7.

In India we usually talk about population growth of human beings. Population also increases of animals and other organisms too. In overseas countries attention is also paid to changes in the population on organisms.

The moose population as well as other organisms have the ability to grow exponentially. The reflect growth of moose and wolf population on Isle Royal National Park from 1900-1985. Moose first appeared on the island shortly after 1900. By 1929, the moose population had grown to about 5000 and had exhausted the easily available food supply. A catastrophic die-back occurred during the 1930, when 97 percent moose died. The appearance of wolves after 2940 at first reduced the moose population but then established a dynamic equilibrium around what appears to be the carrying capacity.

Reasons for exponential growth - the main reasons with regard to the growth of population with given appropriate environmental are as follow:

(i) Biotic Potential - Biotic potential means maximum reproductive rate of an organism. There are many other organisms which have fastest reproducing capacity compared to moose and humans. Housefly has fastest reproducing rate. Each female fly lays an average of 120 eggs in each generation. The eggs hatch and mature into sexually active results and lay their own eggs in fifty six days. In one year if all its offspring survived long enough to reproduce, a single female could be the ancestor of 5-6 trillion flies. The result is astounding.

(ii) Biological Reproduction - Experience of biologists reveal that each cow moose reaches sexual maturity at the age of three or four. If she remains healthy and has enough to eat and survive it can produce one to three calves each year for the next eight to ten year. This means that each female moose can produce as many as 30 calfs during her life time. If half of those calves are female then the population will increase at a 14 to 35 percent annual growth rate. But within a very short time, the numbers began to increase. The growth curve produced by this constant rate of unfettered growth is called a J curve because of its shape. Moose population growing at 35 percent doubles every two year. You can also apply this rate to calculate doubling time of human population. Countries growing at 4 percent per year will double their population in 17-5 years.

Growth of Population

Growth of population is a function of birth rate and death rates. Number of persons born per thousand people during a certain period of time in a country is called birth rates. Growth rates is the rate at which the population grow in a country during a certain period. It is the difference of the birth rate and death rate during the period. If birth rates exceed death rates population will grow and it is called growth of population.

During the early period of Human history, population growth was extremely slow because death rates were higher than birth rates. Reasons for high death rates are as follow:

a. Absence of medical aid.

b. Hostile environment and prevalence of fatal diseases.

c. Scarcity of food and the cave man had to struggle hard for survival.

This happened during 1821. It was called period of great divide. Discovery of agriculture lead to more settled life. Famines and deaths due to scarcity of food registered a decline. All this lead to higher survival rate. Population did increase but slowly. People migrated to new areas and more land was brought under-cultivation. These developments made world population 300 million. It happened at the beginning of the Christian era.

Industrial revolution marked the next important stage in the growth of human population followed by advancement in medical science. Advances in the field of medicine resulted in the control of plague, smallpox, malaria and other deadly diseases.

Protected water supply, sanitation of other preventive measures further reduced death rates. Not only that, settlements encouraged population growth. This happened with the opening up of lands in North-America. South America and Australia. So population increased by immigration as well.

Factors Affecting variation in a Population

The main factors that affect distribution of population are natural resources, climate conditions, soils, cultural factors, age of human settlement, industrial development and means of transport and communication. The factors affecting distribution of population are described below one by one in a nutshell manner:

(i) Cultural factors - Cultural factors like the modes of life of the people also affect the density of population. The density of population is low in equatorial forests and deserts because nomadic tribes live in regions are often found engaged in food gathering to support their families. The regions inhabited by pastoral nomads also have low density because they migrate with their animals is search of pastures.

(ii) Age of human settlement - Australia and USA have low density of population. Human habitation in these countries is of recent origin compared to river valley of the ancient period. Besides this, these countries also practise strict immigration procedures. This system keep population density far below the total potential.

(iii) Means of Transport and Communication - Highly development means of air, surface and water add to the density of population in a particular region because there the people can manage and carry on agricultural, commercial as well as industrial activities very safely and swiftly.

(iv) Soils - Rich and fertile soil always help in the growth and development of agricultural activities with substantially high yield per hectare. This is so because fertile lands produce a variety of crops to support dense population. Son on that East and South Asian Region and northern plains of India are densely populated.

(v) Climatic Conditions - Climate also influence growth of population. Moderate climate leads to the growth of population because it is pleasant and refreshing. In these regions density of population is high. On the other hand people to not like to live in too hot and too cold climates. This is the reason that Canada is sparsely populated. Average density of population in Canada is 2.5 per sq. km. metre. Similarly in India, Rajasthan, Deccan hills, plateaus and the Himalayan regions are sparsely populated.

(vi) Natural Resources - The region having rich wealth of natural resources like coal, water, minerals and forest wealth favour the growth of population. These resources help in the development of trade, commerce and industry. This promote cluster of population and thereby increase density of population. This is the reason that North-West Europe, Bihar and West Bengal in India have high density of population.

Distribution of Population

The total of the world have crossed about 5.8 billion in July 1996 and in the year 2000 it was estimated to be 6055 million. Annual addition to the world population in the next decade will average 97 million. All this population growth will be in Africa, Asia and Latin America. Over half of it will be in Africa and South Asia alone. The details of this uneven distribution is given below:

(i) the countries such as China, India, Russia, Indonesia, Brazil, Japan, Nigeria, Bangladesh and Pakistan are the top ten nations in order of having more than 100 million persons.

(ii) The continent of Asia including Russia has nearly two-thirds of the world's population.

(iii) More than 90% of the population live in the Northern Hemisphere.

Density and Regions of High density of Population

Density of Population which means population per sq. km is necessary to be studied because absolute number of people living in a country do not provide any indication of the impact of population on the land and its resources.

The average density of population for the world as a whole is about 38 person per sq- km according to 1992 estimates, assuming that the entire population is spread uniformly over the land area. The actual density of population varies from region to region. This variation depends upon environmental conditions.

Variation in density of population is classified broadly in three parts as described below:

a. Low density Regions - The density of population in these regions is less than 50 persons per sq. km. Over three-fourth of world's area has density below 50. Low population density regions are:

(i) Tropical deserts like Sahara and Western Australia.

(ii) Equatorial rain forest regions such as Amazon and Zaire.

(iii) High mountains and dissected plateaus with rugged relief.

(iv) Polar regions such as Antarctica and Greenland.

b. Moderate Density Region - These are those areas where 50-100 persons live per Sq. Km. These are regions of extensive agriculture and pastoral activities. Industrial development in these regions is quite more. These regions are:

(i) Coastal low-lands in middle latitudes regions of South America and South Africa.

(ii) European Russia.

(iii) Central and Southern Europe.

(iv) River valleys of South-East Asia.

c. Regions of high density - Areas where over 100 persons are found living per sq. km. are called region of high density. Broadly there are four major regions of high density of population. These regions are:

(i) North-Eastern United States comprising, the region between the Great Lakes and Atlantic coast.

(ii) North-West Europe includes Great Britain, France, Belgium. the Netherlands, Denmark and Germany.

(iii) Southern Asia comprise India, Pakistan and Bangladesh.

(iv) Eastern Asia which include China, Japan and Korea.

Population explosion

Baby boom is called population explosion. There is sudden and fast increase in the birth rate of babies. India is caught in the grip of population explosion. Here birth rates are higher than death rates inspite of family planning measures. The population in the organism tends to increase as long as its environment allows. Thus, most populations are in a state of dynamic equilibrium which is influenced by the local environment and by the interaction of the organisms with other populations in the community. Malthus concluded that human populations tend to grow until they exhaust their resources and become subject to famine, disease or war. Mathusian theory of population appeared to be correct in case of India because he said population has tendency to be doubled after 25 years. Malthus painted a bleak picture of the countries if the population continue to rise at such as learning speed. He opined that there will be imbalance between population growth and the food supply. Under these circumstances, T.R. Malthus predicted cannibalism in the future. In order to control the population growth, he suggested preventive and positive checks. These are as follows:

a. Positive checks - These exercise their influence on the growth of population by increasing the death rate. These checks are natural catastrophics, famines, droughts, floods, and spread of epidemics etc.

Malthus further stressed that the nation should immediately adopt preventive checks to control the population otherwise nature shall interfere through its untold miseries, which has been discussed in the preceding paragraph.

Malthus's theory of population to some extent is found applicable in under-developed, developing countries amongst which India is one. A critical study of Indian economy reveals that population of this country has almost doubled after a period of 25 years. The population of India in 1950-57 was 34.7 crores and now it stands more than 1`00 crores. India is having food problem due to faster growth in population. Increasing rate of population has reduced the ratio between man and land. As described by Malthus, we are using family planning methods to control our population,

Today India is the second populous country in the world; the first is China. Amongst States, Population-wise U.P. is the biggest state, Kolkata is the most population city but now Mumbai has been declared most .populous city,. Being the second populous country in the world according to 1991 census. India constitute about 16 percent of the total population of the world. In other words, every seventh person on the earth is an Indian. India on the other hand, has got only 2.4 percent of the total land area in the world. China with about 20 percent of the world's population has nearly 7 percent of the land area, the U.S.A. maintains only 6 percent of the total world population and about 7 percent of the total area.

India is overpopulated country. Thomas Robert Malthus in his book Essay on the Principles of Population mentioned that human population in every country increase faster than the means of subsistence. Though the population has increased in every country of the world but in India the growth rate of population is quite high., In 1951 the population of India was 34.7 crore which increased to 54.7 crore in 1971. Today India's population is approximately more than 100 crores and 20 lacks.

Causes of Population Growth - Population growth is a function of birth rates and death rates. If BRs increases. In India BRs increase faster than Drs population increases. In India BRs are more than Drs for which following are the main reasons.

(i) Children in rural areas are considered asset rather than liability and 75% of India s population live in villages.

(ii) Wide spread poverty and no source of entertainment.

(iii) The joint family system further provide spurt to population growth.

(iv) the population of women in the age of child bearing age in 15.49 years and 84.53 percent of the total female population falls in the reproducing age.

(v) The average age of women in India for bearing conception is low.

(vi) Marriage is universal in India which means every girl get married sooner or later.

2. Preventive checks - These checks exercise influence on the growth of population by bringing down the birth rates. These checks are used as family planning methods such as contraceptives, late marriage and self-restraint.

Demographic transition theory of population

In the theories of population it is mentioned that population increases fast but its rate of growth falls with the industrial development. Improved medical facilities also ensure low death rates. The historical sequence led to demographers to ask whether this process will continue with industrialization. The theory to demographic transition explains that a stage is arrived at where population tends to stabilize once a certain level of economic development is achieved. There are three stages of demographic transition. These are as follow:

Stage one - In this stage both death rates and birth rates are high. Due to this reason population growth remain fairly stable. This stage is found in traditional societies.

Stage two - In this stage birth rates remains high but

death rates fall due to availability of medical facilities and better living standards. This stage is found in all development societies. Population increases quite fast in this stage.

Stage three - In this stage both death rates and birth rates starts declining and settle at a low level.

Stage four - Population growth rage gradually declines toward zero and remains fairly stable. This stage is found in advanced industrial societies.

The transition has been completed in Europe, Japan and North America. India is at the second stage of demographic transition. The problem of recognizing demographic theory of transition is that it is only a bypothesis and every society do not behave as explained in this theory . Japanese method of controlling population after world war second has been abortion. Secondly decline in birth rates is closely connected to socio-economic development per capita income and educational level.

Optimum Theory of Population

The exponents of this theory comment that population growth as described by Malthus is not always bad. Increase in population helped the economic growth of Western European countries rather than hindering it. This theory of population tries to explain that the size of population which is the best for a country of a particular time. The theory conceives the right number of population relative to the amount of actually developed natural resources, availability of capital and technology. Optimum population is that level of population at which output per capita for real income per head is the highest. If the population happens to be less than optimum level, it is underpopulated and if this is more than optimum level, it is over populated.

Let us assume that the stock of natural resources, capital and technology is given and remains fixed at a given time. Now also assume that the population at initial stages is found less but begins to increased. With the increase in population, the labour supply also increase. With small population and large resources, the production appears to be less because resources are not fully and optimally utilized. So there will be three advantages if population increase:

1. Food problem shall be caused due to over population.

2. Open and disguised unemployment.

3. Low standard of living. So when resources are not fully utilized due to less man power because manpower resources are short then it is the resources then it is a symptom of economic development than retardation.

Dalton has given a formula through which it can be judged to what extent actual population of a country deviates from the optimum population. The degree of deviation is called maladjustment.

Maslthusian theory of population

Robert Malthus, an English clergyman put forward the theory of population in his boom An Essay on The Principle of Population in 1798. He was impressed by the unlimited urge of sex to produce children and limited power of earth to produce food. His explanation has been reduced to the following points:

(i) Law of diminishing returns apply to agriculture. Population get food from agriculture.

(ii) Population has a tendency to be doubled after a period of 25 years.

(iii) There are two ways viz., positive and preventive checks which can keep the level of population checked.

(iv) Population always increases when the means of subsistence increase.

(v) Human population increases faster than food production. Population increases in geometric progression and food production in arithmetic ratio. Geometric progression is 2, 4, 6, 16, 32, 64, 128, and arithmetic progression is 1, 2, 4, 6, 8, 10.\

Malthus painted a bleak picture of the countries if the population continue to rise at such an alarming speed. He opined that there will be an imbalance between population growth and the food supply. Under these circumstances, T.R. Malthus predicted cannibalism in the future. In order to control the population growth, he suggested preventive and positive checks. These are as follows:

(i) Positive checks - These exercise their influence on the growth of population by increasing the death rate. These checks are natural catastrophes, famines, droughts, floods, and spread of epidemics etc.

b. Preventive checks - These checks exercise influence on the growth of population by bringing down the birth rates. These checks are use of family planning methods such as contraceptives, late marriage and self-restraint.

Malthus further stressed that the nation should immediately adopt preventive checks to control the population otherwise nature shall interfere through its untold miseries, which has been discussed in the preceding paragraph.

A storm of controversy was raised over Malthusan theory of population with the following pints of criticism.

(i) Malthus gave no exact proof about his assertion that population increases in geometric progression and food supply in arithmetic progression.

(ii) Malthus held the view that with the increase in subsistence, food supply will increase population. But in western countries, there has been control over birth rates in-spite of increase in subsistence resources.

(iii) Malthusian concept of population is believed to be governed by the law of diminishing returns as applied to agriculture. Advancement in technical know-how and application of technology is agriculture has failed this law.

(iv) In western countries population has failed to grow as rapidly as predicted by Malthus. As a result the living standards of people have increased.

Family Welfare Programme

Most population experts believe that less developed nations must initiate family planning or welfare program in order to check population growth. This programme allows couples to determine the number and spacing of offsprings. For less developed countries family planning programme can accelerate the decline in birth that occurs slowly through economic growth and the demographic transition. Some researchers believe that family planning program are responsible for a 40% to 50% decline in fertility in these countries.

It means smaller size of family that is one to two children may be both male or females. The family should by health accept car and not baby. Family planning programme has been initiated by health department of the Government. Huge amount of money is been spent by the Government of India because India is second populous country in the world only second to China. We have already crossed one hundred crores people by now. More than 30-40 crores are married couple in our country. If we observe the principal of the family planning than by the end of a few years the population of this country shall be more than China. Family welfare programme which is a need of the day stress the following recommendations.

1. Providing incentive to those who can have small size of family.

2. Opening more and more family welfare centre in rural planning measures.

3. Spreading and advertising about the use of family planning measures.

4. Children may not be considered gift of nature but result of mating either sexes.

5. Disseminating education in rural areas in such a style that they prefer small size of family.

6. Spreading family planning programmes in rural areas of Indian society.

7. Educating woman for having smaller size of family.

8. Banning universal marriage.

9. Fixing marriage age of both male and female. It is 18 years for female and 21 years for male. In urban areas this age is on the hire side.

10. Late marriage.

11. Sterilization both bisectomy and vasectomy.

12. Using contraceptives at the time of male-female mating.

13. Having healthy children and not bringing them up in a proper standard is violence.

14. Child by choice and not by accident.

15. One or two children to a family.

If above measures are adopted effectively than

sufficient economic welfare can be exercised to India society.

Future of Human Population

World population is estimated to have grown at the annual rate of 1.4% during the year 1990-2000. There is net increase of three humans per second in the world population because in every second, on average, five children are born somewhere on earth and in that same second, two other people die. In other words BRs are five per second and Draw are two on earth.

The difference between BRs and DRs it three humans. If you multiply this out, you will find we are growing at a rate of about 11,000 per hour, 2,65,000 per day, or almost 100 million more people per month. By 1995, the world population had reached about 5.7 billion. Many people worry that overpopulation will cause resources depletion and environmental degradation that threatened the ecological life. Others believe that human population is useful because it provides technology and enterprise that extend the world's carrying capacity which allows us to overcome any problem we encounter.

A larger population means a larger workforce, more genius, more ideas about what to do. Along with every new mouth comes a pair of hands.

Environment and Human Health

Health means disease free body. The state of the people does not depend on the number of doctors and hospitals, but also on clean environment. An unclean environment can cause a number of health problems because ecological changes have a direct impact on human health. There are certain changes in the human environment that increase the incidence of many diseases. Such changes include socio-economic and cultural changes leading to stress of many kinds of human health.

For instance, urbanization and industrial stress, population stress and others ultimately bring about changes in the human environment, making it a paradise for infectious agents. Nearly 80% of the world's diseases, particularly in developing world can be linked with water. In our country there are 14 river systems. About 90% of pollution load in river system is due to human waste. For instance during Kumbh coliform increases from 300/100 ml to about 165,000/00 ml. In our country Kerala has much lower mortality rates that even the nation average. The reason are high literacy in women and better environmental sanitation. In Rajasthan the situation is very bad. Nearly 8000 villages in Rajasthan and Central India are affected with Guinea worm disease. Return of malaria and filaria and kalabaazar in Japan is also due do to adverse environment conditions.

Keeping the dangerous effects of bed environment following points must be kept in view:

1. Health care should not be biased towards cities. There should be proper care in villages also.

2. Disease control should not be taken in isolation. There should be cooperation of doctors, health educators, social anthropologists, media men, engineers, socio-economics and others.

3. Habitat management is very vital.

5. Future impact assessment of all development projects must include health aspect as well.

Contaminated water is the main cause of bad human health. The potable water contaminated with faeces is the chief cause of some important diseases of man. The enteric diseases are transmitted mainly by swallowing food or drinking contaminated with faeces. Typhoid fever, dysentery, cholera and other enteric diseases are caused by consumption of contaminated water. Some common human bacterial disease transmitted by factually contaminates are water:

Sanitation Bulletin: An emergency sanitation bulletin should include the following:

1. Milk and Food - Use only pasteurized milk. When the safety of milk is in doubt, or when only raw milk is available, bring the milk to a boil and cool in a clean container. If boiling is not possible, canned or powdered milk mixed with safe water can be used.

Cook all foods thoroughly. Frozen foods that have been thawed should be discarded if not consumed immediately or kept refrigerated at 450 F for lower. Avoid creamed dishes, hash, puddings, meat, poultry, and pre-prepared salads unless refrigeration can be assured.

Utensils can be used if clean and not exposed to flood waters. Raw foods exposed to flood waters should be avoided unless cooked because of possible contamination. If raw foods necessary to be used, clean thoroughly and soak in chlorinated water at least for 5 minutes.

Destroy the contents foods in tins where selling, leaking, rusting, or serious denting is visible. Wash off and immerse all hermetically sealed cans in good condition in a bleach disinfecting solution for at least 5 min prior to use in the same manner as above.

Clean and then disinfect all utensils and equipment prior to use in the same manner as above. Silfer should be immersed in boiling water for at least 2 min. Use single-service paper or plastic dishes, cups, and utensils where possible.

2. Safety Precautions - Entering into Damaged Buildings- Proceed with caution. If you have any doubts about the structural safety of a building, seek professional advice or assistance before going in. Check for visible signs of buckled walls, loose bricks, cracks, or shifting of foundation.

3. Household Cleanup Procedure - Cellars- If possible, wait for the groundwater level top drop below floor level. Otherwise, drain or pump water from flooded cellars, being careful to avoid the collapse of walls caused by the pressure of the water-saturated ground around the basement. Postpone pumping if groundwater level is high. Wash down the walls, floors, and other areas exposed to floodwater; keep windows and doors open for ventilation.

Disinfect the washed area by applying a solution of bleach with a broom or stiff brush. Allow the solution to remain in contact for approximately 10 min. The solution is prepared by adding 2 oz of bleach to 5 gal of water. Rinse with cold water as soon as possible to minimize staining. Any commercially available disinfectant can be used.

(i) Furniture - Clean and then wash metals, plastic, and leather surfaces with mild soap and water and wipe dry immediately. Some upholstery may be washed on the surface with soap and water and wiped dry. Expose to open air and sunshine.

(ii) Floor Coverings - Flush rungs and carpets with house and squeegee, the wash with lukewarm water containing a detergent. Rinse and dry in sun.

4. Refuse Collection, Storage, and Disposal - Drain and warp garbage in several layers of newspaper to absorb moisture. Store garbage in watertight, rodent-and insect-proof containers having a tight-fitting cover. Use

plastic linear if available. Bury garbage when necessary in a hole deep enough to provide for at least 18 in. of dirt cover and located at least 100 ft from any well of surface water.

Store all other refuse in a convenient place until normal collection services are provided, or take to the nearest designated place established by local authorities. Bury all dead animals as quickly as possible at a distance of a least 100 ft from any well or surface water and cover to a depth of at least 2 ft.

5. Insect and Rodent Control - To prevent breeding, store and dispose of garbage, sewage, and dead animals according to instructions. Drain standing water where ever possible, including containers of stagnant water. Treat standing water with kerosene, fuel oil, or commercially available larvicide during mosquito breeding season; apply 3 oz per 100 ft of water surface or in accordance with the manufacturer's instructions.

To kill rats, when trapping proves unsatisfactory, use commercially available quick kill poisons- such as zinc phosphide or red squill. Follow directions on the label. USE WITH CARE! Keep these poisons out of the reach of children and away from animals. Store in a locked cabinet.

6. Sewage and Excreta Disposal - When flush toilets cannot be used, use a small covered watertight bucket with a removable plastic liner to receive body wastes. Use a small amount of chlorine bleach, chlorinated lime, or any commercially available preparation which disinfects and deodorizes; follow manufacture's instructions. Remove the plastic liner as necessary; tie and place into a large watertight, rodent- and insect-proof garbage can with a tight cover for temporary storage. The wastes may be disposed with the garbage when

normal collection is restored, or by burial in a deep hole with at least 18 in. of dirt cover located at a distance of not less than 100 ft from any well or surface waters, or when and where feasible, by emptying the contents into the flush and carefully disposing of the plastic bag with the garbage.

Use the temporary public toilets established in the disaster area; they may be located over sewer manholes.

7. To Disinfect Your Well - First, pump well until water is clear; then:

(i) Pump heavily chlorinated water to waste, away from grass and shrubbery.

(ii) Repeat steps 1 through 3 to ensure that the chlorinated water has circulated through the entire toilet flushing.

(iii) If possible, reopen one tap and recirculate water back into well by means of a garden hose for a least 1 hour, then close tap.

(iv) Pump until chlorine odour appears at all taps, then close taps.

(v) Pur solution into well though top of casing or use vent pipe where available.

(vi) Mix one of liquid bleach in 5 gal water.

8. Drinking Water - Use Only Safe Water - Assume all sources are unsafe for use until approved by your local health department. Meanwhile, use health department-approved bottled water, or water distributed by health department-approved tank truck.

Or, disinfect water using a household liquid bleach disinfectant containing at least 5 percent available chlorine by weight. To one gallon. Of clear water, mix 6 to 8 drops of bleach; let the solution stand for 30 min before drinking. If the water is cloudy and contains particles, allow the particles to settle. Pour off the clear water into a separate container and add to it double the amount of bleach mentioned above; or heat water and a rapid boil for 2 min; or disinfect water using purified water using purification tablets. Carbonated beverages are satisfactory.

Value of Education

The environmental scenario of India is wide and diversified. Climatically, geographically, geologically, socially, economically and ethicality this country is highly diverse. It is, therefore suggested that while imparting environmental education special attention must be paid to school going children and women because they alone constitute about 50% of the population. They should be made aware of health, nutrition, sanitation, hygiene, water and food contamination. The main objectives of environmental education should be improving the quality of environmental, creating an awareness among people on environmental problems and conservation, creating such an atmosphere so that people find themselves fit enough to participate in decision making so that development programmes are presented after evaluation.

Role of I.T. in Environment and Human Health

Information technology is the need of the day. Computers, satellites and other technological developments have enabled environmentalists to gather informations and forecast the disasters much quickly with I.T. The engineer or environmentalists can gather world wide information on his table and make use of the same as per his requirements. Through GIS and GPS, a large amount of information along with photographs are easily accessible.

ENVIS, set up by the Ministry in 1982 as a decentral information system network continued its activities in collection, storage, retrieval and dissemination of environmental information to decision makers, policy planners, scientists, engineers, environmentalists, research workers and the general public all over the country. The ENVIS network besides the focal point in the ministry presently consists of 13 subject oriented centres, known as ENVIS Centres, set up in the various institutions/organizations of the country in the priority areas of environment like pollution control, toxic chemicals, energy and environment environmentally sound and appropriate technology, bio-degradation of wastes, deserification, estuary, mangroves, corals and lagoons, media and environment etc.

ENVIS has been designated as National Focal Point and Regional Service Centre for South Asia sub-regional countries by INFOTERA of United Nations Environment Programme. It caters to the environmental information needs more than 500 Indian Protectial Sources in the international directory of UNEP of providing information in the field of environment and its related areas.

Women and Child Welfare

Two-thirds of the world is done by women, whereas they own less than 7% of world's wealth. Women play a great role in unbringing family. Child welfare is situations which comes after bodyhood. She has the responsibility of child welfare. Fastest growth takes place during the first 2 years of life and again between 10 and 15 years for girls and 12 to 17 years for boys. They need the best food but many do not have. During this time they develop a lot of physically, mentally and lots of skills and ideas. The vocabulary develops from 0 and 1 year ago to about 2500 words at 6 years. Ideas about right and wrong develop through the example and teaching of adults. Through games one learns the importance of rules to that the game of life runs smoothly. Simultaneously, friends become very important and influential too.

In 1971 member countries of United Nations agreed on the declaration of rights of the child. A treaty was brought out by united nations of the Convention on the Rights of the Child", which was published in 1919. Out of 159 countries 100 countries accepted the rights of the children and the role of women towards child welfare. Following are the highlights of this convention.

1. Education which play facilities must be provided so that they grow well on all fronts.

2. Every effort should be made for medical care.

3. Clean and ventilated housing should be arranged for them.

4. Children should be provided nutritious food.

5. All children have the right to love understating.

Remote Sensing

Remote sensing is an art by which information can be acquired about an object or a surface feature from measurement at a distance without coming in physical contact with object or feature of interest. This information by noncontact methods is acquired usually from airplanes or satellites. It is one of the areas of rapidly advancing technology, very useful in ecosystem and environmental management. Photography from the ground or aerial platforms is a conventual form of remote sensing being used by foresters, geologists and geographers. Information on photographs or other images can be evaluated qualitatively and quantitatively. The method of concluding relationships by visual examination is called photointerpretation whereas the procedure of making photographic measurement is called photogrammetry.

Role of remote sensing in ecological research is enormous. It contributes to inventory and mapping of resources, quantifying the environment, describing the flow the matter and energy in the ecosystem, and evaluating change and alternative solutions for management of ecosystem. All these together help in environment management.

Remote sensing is proved very useful in soil and agriculture, town and country planning, detecting depiction of forest cover mapping etc. Remote sensing provides information about various agricultural resources that may have effect of the crop production. Many countries including USA and USSR have got remote satellite programme for land resources survey, environment impact assessment, weather forecasting and ocean studies. The proper know-how and technology is also developing in India by National Remote Sensing Agency assisted by Indian Institute of Remote Sensing, Dehradun. This Institute impart training to scientists and engineers in the field of remote sensing and photointerpretation since 1086. India launched many remote sensing satellites such as Bhaskra II and India satellite for earth imageries with coverage of sufficient radius. The use of this information can be made to exercise environment management and planning.

Global Positioning System

The Global Positioning System is an excellent example of information system based on modern radio and satellite systems. GPS demonstrates the technology and capabilities that were inconceivable a short time ago. GPS incorporate diverse types of information technology and it serves a broad range of utility guiding smart bombs or missiles to locating fish in water.

The basic function of GPS is to determine the position of GRS receiver anywhere on or above the earth in terms of latitude, longitude and elevation above sea level, with a clear view sky so that signals can be received from at least three GPS satellite at different locations in the sky. Once the position in the unit are determined by the receiver, the processor may then convert that location something useful to the user.

If the signals are from three satellites the GPS can determine the location anywhere in the surface. With signals from four or more satellites, the altitude can also be determined so that position in 3 dimensions may be determined. GPS is an excellent example of generating new products and capabilities. For a environmentalists GPS can be of great use. It can be used for many purposes, which could not be covered by remote sensing.

Geographic Information System

The management of resources need detailed inventory of all natural resources- abiotic and biotic. GIS study is very beneficial in this direction. The information so collected and compiled is quite useful to future generation. Help of computer is necessary to avoid negative consequences while compiling information because information collected manually shall become outdated because the resources would disappear as they get spoiled in the process of environmental degradation with passage of time. With the introduction of commercial geographic system, it has been possible to creatively analyse and interpret the data, eliminating the slow, laborious methods of data handing. Both CAC and CAD on the one hand and GIS on the other are equally beneficial. The GIS is an important tool for converting the spatial data into information which can be utilized to support the decisions for the intelligent use of earth resources and to manage the man made environment. Contents of GIS are given below:

(i) One can develop a small database of his own against which he can compare the test results.

(ii) It is possible to preserve the majority of the attributes data.

(iii) The conversion between rester and vector is quite common.

(iv) GIS use raster or a vector format to represent location.

(v) It provides specific location data with location, attribute and time.

Need for Planning Future

A review of the above description suffice to bring out the necessity of a direct attack on the problem that should aim at a rapid reduction in the birth rate. Based on the experience of western countries economic development and industrializations are quite often stressed as the best contraceptive. Besides this there are a number of remedies stated below if exercised effectively can check population in planning future.

Family planning is limiting the number of children to the born to a couple to one or two, More intensive research in family methods and practices, Creating grater awareness of the opportunities for legal fermentation of pregnancies, Setting up of family planning centres to make available the various related family planning, Using various means of communication to persuade people to through different outlets in urban, semi-urban and rural areas, Determining the spacing of children.

Density of Population

The density of population is calculated as ratio of the number of persons per sq. km. of land area. According to the 19912 census, the density of population in the country is 267. In Burma it is 50, Japan 316. One can know 3 interstate variation i.e., 6,319 but Arunachal Pradesh has only 10. The factors governing density of population are:

1. Rainfall irrigation facilities and agricultural development.

2. Better climate.

3. Trade and Business.

4. Industrial advancement.

Ecosystem Case Study/Environmental Case Study

Kolleru lake, a wetland located in Andhra Pradesh, is one of India's largest freshwater lakes and a bird sanctuary. Between 1985 and 1087 a study identified more than 185 bird species in the area, and new species continue to be found. The lake harbours four rare bird species and 12 endangered ones. It was declared a protected area of pelicans- about 8,000 pelicans came in 1961- but since 1972, none have visited. But today, economic and industrial development, expanding fisheries and pollution threatens Kolleru.

1. Ecological degradation - Pressure on the lake has led to proliferating weeds and fewer visiting birds. The catchment area has shrunk. This has led to eutrophication, loss of drinking water and declining fish catechus. Obstructions on the lake's periphery lead to flooded agricultural land even during normal rainfall.

The shape and area of the lake is difficult to assess because floods submerge large areas. The water spread varies from 135 sq km at + most level to 901 sq km at +10 mst. The average depth of the lake varies from 0.5 metres to two metres: the lake is silting gradually, raising the bed level.

Sewage inflow from the towns of Eluru, Fudivada and even Vijayawada and industrial effluents, pesticides and fertilizers from the Krishna-Godavari delta region contaminate the lake. Eleven major industries release about 7.2 million liters of effluents into the lake every day. An Andhra Pradesh Pollution Control Board report states that more than 17,000 tones of fertilizer wash enters the lake annually. Studies have shown the presence of organic pollutants in lake sediment and the weeds to growing fast. The sewage and discharge from factories have also affected the growth of water-bone organisms that the fish consume.

An ecological survey conducted in 1978 with financial assistance from CSIR, found no evidence of algal blooms and no significant contrast between the concentration of dissolved oxygen at the surface of the lake and at the bottom. However, in 1980 by the Administrative Staff College of India stated that the weed problem in the lake was caused by high levels of pollution. In 1972, state pollution board member warned that and dead zone would be created if pollutants were not checked.

2. Fishing - The roads and bridges that had come up with agricultural development in the area, along with an increased demand for fish, especially from Kolkata, created a new and vast market for fish by 1978. Pisciculture suddenly became profitable. By 1984, 5,000 acres of government lake bed land were converted to fish tanks under corporative societies. The government did well. Financial help was provided by pledging the

petty lands to cooperative banks in order to construct fish tanks. Our department provided technical help and training. Unfortunately, due to poor maintenance and low productivity of these tanks, fishefolk could not repay the loans. The government auctioned the tanks to private investors.

3. The led of Kolleru Lake has been used for ages to cultivate yerra vari, a native paddy variety. Since 1940 the cultivated area increased as the British government granted pattas on payment of market value for the land. 1954 on, the government initiated cooperative farming and organised 93 farming societies on 2.1 million acres of the lake bed. By 1969, almost the entire lake came under cultivation with huge bunds constructed to keep water out. The native paddy was replaced with shorter, high-yielding varieties requiring more fertilizers and pesticides.

Effects of Population

Reduction in death rates per 1000 and not controlling of birth rates to desirable limits has led to what has been described as population explosion. Rapid increase in population is threat to the nation's development. Population growth has nullified the efforts and improvement in the quality of life. Unlike developed countries, in developing countries like India where labour is abudant and capital scarce population growth is liability rather than an asset. Discussion to this reflect effect of population on nation's growth rate has been shown as follows:

(i) Illiteracy and ignorance shall prevail everywhere in India in the near future due to rapid growth of population according to available estimates. Newly born children will need one or two more fully equipped schools in each district every week.

(ii) At present food production in the country is about 180 million tonnes which is just sufficient to maintain the present level of population at the existing standard of nutrition. Any further increase in population, unless met with an equal proportional increase in food supply, will certainly lead to the shortage of food which will lead to famine, resilience, malnutrition, epidemics etc.

(iii) The pressure of population on land which is already high, is rising further as a consequence of growing population.

(iv) Growing population adds to the problem of under-utilization of labour.

(v) The growth in numbers has itself contribute to environmental damage, especially through reduction of soil cover and consequent flooding.

(vi) Population has adversely affected the growth of per-capita product.

(vii) High fertility rate and declining mortality in younger and old age group, the population is acquiring an increasing proportion of people in the non-working age groups relative to those of working age.

Declining Death Rates

India is fast in closing the gap with the rest of the world with regard to death rates. Death rate in the world in 1991 11 per 1000. This means 1.1% Following are the reasons leading to fall in death rates:

1. Food shortage which used to cause famine deaths have been overpowered.

2. Provision of better maternity and post-natal care has helped to bring down the death of infants.

3. There is growth of sanitation and cleanliness which help to reduce the incidence of morality.

4. Plague, Smallpox, Typhoid which used to finish villages in past have now been controlled with the use of antibiotics and other life-saving medicines.

5. Technology with regard to diseases control.

2

COMMUNITY AND ENVIRONMENTAL EDUCATION

All over the world, drinking water and sanitation projects are being viewed as a base for promoting public health. In India, the coverage of sanitation services are poor, to achieve the Million Development Goals (MDG) there appears to be insufficient resource allocation to the sector. Hence, under sector reforms the government replaces heavy subsidized programme with demand driven and community led approach, where community share/recovery of partial capital cost, to reduce burden on state exchequer. This study investigated how and to what extent this shift help in achieving the set objective by choosing a similar project implemented in Karnataka state with the following objectives:

- To examine community contribution for environmental sanitation.
- To examine impact of community participation in expected outcomes.
- To examine the contribution by different stakeholders during implementation of project

Approach and Methodology

To test the objectives, a multiple sampling method was used to select the villages and households. The result in analysis shows that the communities were unable to contribute for the capital cost. As a result, there was inequity in sharing the benefits of the project among the community. Hence, it necessitates redesigning the project with specific role for each actor in the project.

Introduction

In recent years, there has been increasing recognition of the limitations of the capacity of government agencies to effectively manage the rapidly increasing multiplicity of development programmes. As a substitute the governments have been assigning greater roles to the Non-Government Organisations (NGO), private sector and communities for greater participation. This shift in approach was observed in various sector including drinking water and sanitation services, to increase efficiency, for cost recovery and sustainability of the schemes/assets, this has been experienced in many countries. In fact, the drinking water and sanitation sector received attention to promote public health in turn the productivity of human resources can be enhanced, thus poverty can be reduced.

In India, just 22 per cent of the households have access to sanitation in rural area (NSS 1998; Census 2001). This level needs to rise to 53 per cent by 2015 to meet the MDG target, a huge challenge. There appears to be a shortfall of some Rs 287 billion in the financing needed from 2002-15 (Water Aid 2005). Toward this goal the Government of India is making all the efforts, traditionally relied on high level of subsidies to construct latrines. The high subsidy approach has been now replaced with the principle of demand driven and community led, with the introduction of the Total Sanitation Campaign (TSC) in 1999. The concept of sanitation has now been expanded to include liquid and solid waste disposal, personal, domestic as well as environmental hygiene.

Location of Project

The Karnataka (Federal State) state implemented three major projects supported by multilateral and bilateral agencies, the new approach introduced in TSC was in-built in these projects. This study investigated how and to what extent the paradigm shift took place in demand driven and community led. For investigation, the Karnataka Integrated Rural Water Supply and Sanitation Services (KIRWSS) project was selected, which was implemented during 1994-2002, in 1104 villages (rural settlements), supported by World Bank.. Two sets of villages were selected for investigation, first set of villages were randomly selected ten per cent of the 1104, project villages. Purposively all the nine pilot phase villages spread across the state were selected for investigating impact of project at household level, the households were selected proportionately to total households. Thus, 121 villages (112+9) and 755 households were finally selected and examined. According to methodology (see Appendix), the required secondary and primary data was collected from different agencies and households.

Situation Analysis and Description of Problem

According to the latest statistics in state only 18 per cent of the households have access to latrines and 35 percent for drainage systems (Census 2001). Of the total access to latrines only 9 percent were have pit latrine. The reasons for the low coverage vary; however, the common factors are religious beliefs and culture, the advent of Hinduism and its caste system (Cooper 1997). The community often unable to perceive a direct relationship between the current health profile of household and sanitation services. (Veerashekharappa,

2001; 2004) As the project was based on need and demand driven concept, the primary objective was to create awareness on impact of water and sanitation related diseases, in turn to create demand for project. For this, the Information, Education and Communication (IEC) strategy was adopted. Under the project, the entire investment on water was met by the project funds, while for environmental sanitation 30 per cent of the cost has to be met by the community in the ratio of 70:30. The NGOs, beneficiaries and private sector have participated; the beneficiary was involved in all stages of implementation through Village Water Supply and Sanitation Committees (VWSC). The concept followed in the project was with better resources and community participation; the expected outcomes can be derived. The process and expected outcomes were well defined under the project, which was reproduced here briefly.

Resource

1. Finance
2. Human resources
3. Institution

Process

1. Participation
2. Accountability
3. Transparency

Expected Outcomes

1. Equity
2. Efficiency
3. Sustainability

As mentioned earlier, the community has to share 30 per cent of capital cost of sanitation. To tender water work construction, the VWSC has to mobilize 25 per cent of the 30 per cent, thus the community was made accountable for initiation of the project. Of the total 12 district (provinces) where project implemented, except in two districts, in all other districts the agreed amount was mobilized, but in none of the district the amount was entirely mobilized from the household source. The balance amount has come from non-household sources, such as Gramapanchayat, cooperative societies, contractors, etc. Of these, the role of local leaders, contractors and Gramapanchayats was important in that order, because together they met the biggest chunk of the gap left by the households, which amounted to 41.6 per cent. Of the total selected (112) villages, including four villages, where no contribution has taken place, in total 21 villages the contribution was less than 25 per cent. The major concern in this context was, of the 29 contractors, 45 per cent of them contributed an amount varying from 40 per cent and upwards of the total contribution in the village (in one village 80 per cent). Finally, the total amount mobilized at state level from all sources under community contribution was Rs 130 million as against the expected Rs 300 million (GOK 2003), thus, 43 per cent of the agreed amount was mobilized. Within that, what was actually mobilized from the household was just around 60 million.

As mentioned earlier, 25 per cent of the agreed amount had to be mobilized for tendering villages for water works, but in practice the norms were flouted. The reasons attributed were that the tendering was done on global competition, to achieve the economy of scale and a couple of villages were combined as a slice on cluster basis. In the process, the villages that had not mobilized the agreed amount also tendered on the pretext that community would mobilize the agreed amount later. But, that could not take place at all, thus the accountability on the part of community was relaxed. In pilot villages, the pattern of contribution was more or less same what was observed in large sample. The most identified reason for less number of household contributions was the inefficiency of NGO in convincing the household about the benefits of the project.

Most of the households were under impression that 'none pays' or `nothing happens even if we did not pay'. Thus, the NGO could not make much effort in changing of mindset of the household, which was very crucial for successful implementation of the project. The inefficiency of NGO was turned out be an advantage to contractors in each village. If the households had been convinced on the relation between contaminated water and health profile of the households the situation would have been different. The role of NGO, along with creation of awareness, it has to facilitate VWSC in its various functions, such as facilitated to conduct meetings, provide technical training in operation and maintenance of assets, etc. As per the norms, VWSC meetings were scheduled every fortnight, but, the minutes of meeting book show that in all the villages, in aggregate 43 per cent of meetings held as per schedule. The reasons for not holding meetings as per scheduled was attributed to the factions in village and reconstituting of VWSC in couple of villages, due to change of party in power. It is observed that women and vulnerable section of the community attended the meetings, because of appropriate reservations were made to them while constituting the committee. But, their presence was not felt in meetings.

The VWSC had been authorized to award contract for the construction of sanitation services. In practice, the persons, who played a key role in mobilizing the large amount, took the decision. In each village, the contractor was awarded to the person recommended by them. If any person questions the award of contract was cornered and suppressed with questions like 'Why should you interfere? Are you going to pay for it? Why bother who gets contract. If you want to interfere, why don't you mobilize more? Moreover, when the contractor was willing to contribute on our behalf, we have saved of paying towards our share of capital cost. Thus, there was no community participation and transparency in awarding contract, as there was nexus between elite group of village and contractor.

Thus, though finance (enough funds), human resources (NGO) and organisation (VWSC) was provided in the project as input in. But, it is noticed that there is not equity provision of services, which were biased towards wealthy localities in terms of accessibility and adequacy of water supply. Of the total households, 82 per cent of the wealthy households found the source located within short distance, 32 per cent drew water 40 lpcd and above, whereas on an average, 17 per cent of the poor drew water less than 40 lpcd. The quantity of water obtained was positively associated with the landholding and wealth status. Similarly, there was inequity in provision of sanitation services. As mentioned earlier, sanitation services made available largely on demand based by sharing proportion of total investment cost. As a result there were large variations in the components provided and proposed, considering the mobilization of agreed amount. For instance, the drainage construction was not taken place in one village, in three villages it was partially constructed that too in wealthy locality. The box type drainage was constructed in wealthy locality, while in other localities the L shape drainage was constructed. Of the total selected households, on an average, 17 per cent of them had Individual Household Sanitary Latrine (IHSL), against the targeted 30 per cent by the project. The 17 per cent were a cumulative figure of various other similar programmes, such as Nirmal grama yojana (NGY) implemented in respective villages.

The reasons for low coverage were attributed, for linking sanction of subsidy to the household toilets with community contribution and private household connection (PHC) of water. The beneficiary had to meet at least one of these two conditions to get the benefit. This undue practice had kept away many deserving households from opting for latrines. At the same time, our investigation found a couple of households misusing the facility, with the nexus of officials and key persons within the village, by showing latrines constructed under NGY or self initiated against the KIRWSS and claimed the subsidy.

Lessons Learnt and Alternatives

The above analysis brings out that; the concept of community participation through cost sharing and recovery in development projects had been theoretically established and empirically tested in many parts of the world. The international agencies have been promoting this idea by funding directly and indirectly, countries are accepting these policies to reduce the financial burden on public exchequer of their county.

The lessons learn from this project was, the programme not deep rooted in minds of people, to achieve this, the approach has to be revised. Both the institutions (NGO and VWSC) were ineffective in various aspects, in fact the VWSC was functioning as an extended arm of the gramapanchayat. Thus most of the demerits of that institution have been inherited by this organization. As a result, neither the beneficiary shared the capital cost nor user charges were paid in many villages. In fact, the project helped the key persons in the village to capture benefits, without accountability. Subsequently, most of the villages (except 10 to 12 villages) were transferred to gramapanchayat, the operation and maintenance cost is met largely from state budgetary allocation.

Hence, there is a need to re-examine at the design of similar projects, particularly designing the process, which is most important. Particularly the job of regulator and service provision has to be separated, to achieve equity and efficiency in service provision, if necessary partially privatization can be acceptable by involving Self Help Groups (SHGs) through linking up micro-finance activities. Indeed this approach might be an alternative in future.

HOME HYGIENE AND ENVIRONMENTAL SANITATION: A COUNTRY SITUATION ANALYSIS FOR INDIA

Problems of the environment and of domestic hygiene are always related to poverty of population and the sanitation of settlements. Most cities and towns in developing countries, like India, are characterised by over-crowding, congestion, inadequate water supply and inadequate facilities of disposal of human excreta, waste water and solid wastes. Inadequacy of housing for most urban poor invariably leads to poor home hygiene. Personal and domestic hygiene practices cannot be improved without improving basic amenities, such as water supply, waste water disposal, solid waste management and the problems of human settlements. But even under the prevailing conditions, there is significant scope of improving hygiene practices at home to prevent infection and cross-infection. Unfortunately, in developing countries, public health concerns are usually raised on the institutional setting, such as municipal services, hospitals, environmental sanitation, etc.

There is a reluctance to acknowledge the home as a setting of equal importance along with the public institutions in the chain of disease transmission in the community. Managers of home hygiene and community hygiene must act in unison to optimise return from efforts to promote public health. Current practices and perceptions of domestic and personal hygiene in Indian communities, the existing levels of environmental and peri-domestic sanitation and the 'health risk' these pose will be outlined, as well as the need for an integrated action for improving hygiene behaviour and access to safe water and sanitation.

UNICEF Partners with Community to Initiate a Cleanliness Drive in Relief Camps

UNICEF partnered with the government and NGOs to mobilise communities in relief camps to clean up the open defecation and build toilets for men and women within the camp area. The exercise kicked off from the Nilayathachiamman Kovil, in Nagapattinam the men and older children enthusiastically shovelled the earth to clean up excreta lying scattered in the open. Government workers and 50 volunteers recruited by the NGO SCOPE supported community.

TRAINING THE COMMUNITY AND THE VOLUNTEERS

Community members and volunteers recruited in

Trichy by SCOPE were first trained in the Centre on the need for maintaining environmental sanitation that focused on the prevention of open defecation, provision of appropriate toilets, refuse disposal, and the importance of washing hands with soap after defecation. Soap provided to UNICEF by Rotary and P&G was distributed by the trained volunteers to complete the process.

The Joint Secretary Health, Supriya Sahu requested the UNICEF team to deploy trained volunteers to three camps where sanitation and water is a critical problem. The Water, Environment and Sanitation (WES) officer at Nagapattinam will deploy 15 of these volunteers at each of the Centres.

The roll out plan of the Community Centre sanitation programme will be need-based covering all Centres in the district that do not yet have any toilets. The Joint Secretary will maintain contact with the UNICEF WES officers to co-ordinate the efforts.

Temporary Toilets

Trenches 3 ft. deep, 2 ft. wide and 7 ft. long were dug in the open compound of the temple, enough for 45 people to use the toilets at one time. For privacy, the trenches were covered with coconut thatch "walls". Special women's toilets were created to protect their privacy - a major issue at the camps.

The UNICEF initiative will ensure that all relief camps have sanitation facilities and the community will be in charge of maintaining cleanliness in these temporary structures.

Community takes responsibility to clean up toilets

The 15 members of the Fishermen's committee along with women groups agree to take responsibility to maintain the cleanliness of the toilets and ensure that no body in the camps defecates anywhere in the open but in the toilets.

Women and children began using the toilets minutes after they were constructed. "This is a great relief. We were defecating and eating food in the same grounds. Many of us felt giddy while eating due to the stench from the excreta," says Meenakshi who comes with her child six year old child the newly built toilet.

Community Water and Sanitation Facility: A Cities Alliance Initiative

In February 2003 the United States launched the Community Water and Sanitation Facility as a Global Development Alliance (GDA), public-private partnership, with the multi-donor supported Cities Alliance to expand water and sanitation services in urban slums. The Facility works towards achieving the objectives of Target 11 of the Millennium Development Goal Number 7 to improve water and sanitation and significantly improve the lives of at least 100 million slum dwellers by 2020. The U.S. launched the Facility with seed funding of $2 million and technical assistance from other agency programs.

The Community Water and Sanitation Facility supports local authorities and their partners in working through public-private partnerships to expand water and sanitation services to slum communities, fund community endorsed construction, promote risk-sharing and innovative financing to mobilize funding resources.

Cities Alliance is a donor coalition committed to the vision, "Cities Without Slums" and has 14 contributing members, including Brazil, Canada, France, Italy, Japan, Germany, the Netherlands, Nigeria, Norway, South

Africa, Sweden, the United States, the United Kingdom, the Asian Development Bank, the United Nations Environment Program, UN-Habitat and the World Bank, plus the four major global municipal organizations under United Cities and Local Government (UCGL).

Public-Private Partnerships Are Essential to Bringing Water and Sanitation Services to Slum Communities

USAID-sponsored Community Water and Sanitation Facility (CWSF) activities underscore and promote the importance of public-private partnerships to achieve sustainable provision of water and sanitation services in slum communities. Several USAID country missions are working with partners to broker activities for Facility support.

USAID-Sponsored Projects

USAID/South Africa has worked with the British Department for International Development (DFID) and the Municipal Infrastructure Investment Unit (MIIU), a non-profit corporation funded by the Government of South Africa, to increase access to water and sanitation services for over 1 million disadvantaged persons in five municipalities. The partners are using the CWSF to support water system upgrades, promote local capital investment, and better managed services for the urban poor. USAID has provided use of its innovative Development Credit Authority (DCA) credit instrument to facilitate local lenders making funds available to these municipalities. With an expected leverage of over 36:1, the Facility contribution of $300,000 should generate $16,675,000 in partner contributions. In a related municipal finance partnership, USAID and private sector partner, Evensen Dodge, are assisting the Cities of Tshwane and Durban to obtain access to up to $1 billion over the next three years to finance infrastructure investments such as in water services.

USAID/Mozambique is now supporting local governance programs which may results in CWSF projects in six cities and a proposed UN-HABITAT initiative to address interlinked problems of poor access to safe drinking water and efficient waste management in Quelimane City. The goal being to promote participatory development strategies and implementation of innovative tools resulting in improved water and sanitation solutions.

In India, USAID and Shelter Associates, a community based organization representing slum dwellers in Sangli, constructed community toilet facilities for the use of 25,000 persons. The project objectives were to:

a. Develop community-level toilet construction and management plans in consultation with local government authorities,

b. To assist the municipality in implementing the community-level plans and

c. To begin the preparatory work for a citywide slum upgrading strategy.

The Facility was asked to fund technical assistance and construction materials for public toilets. Added resources were provided by private sector and community partners. Upcoming CWSF efforts are in development, where USAID/India will partner with local stakeholders and the Government of India in Bangalore and Agra to address long-term city wide infrastructure planning, delivery of basic services and test a pilot "willingness-to-pay" demonstration program. The pilot will showcase how low income communities can acquire reliable services when community residents pay for individual connection charges and monthly water and sewerage tariffs. This win-win mechanism has enhanced revenues for the utilities thereby enabling it to improve service quality to slum residents.

USAID/Panama has been working with local stakeholders in Panama City to plan a CWSF program to address public health concerns and bring sewerage lines and treatment facilities to a marginalized community of 12,500 people in the Panama Canal watershed. Through a series of community meetings, participants identified how toddlers and women were disproportionately affected by skin and gastro-intestinal problems due to the lack of sanitation services. In response to the community's needs, USAID through the CWSF is helping to broker a public-private alliance with the Ministry of Health, Panama Canal Authority, the National Institute for Water and Sanitation, the Municipality of Panama, local business and community residents to construct a affordable sewerage system. This partnership will help protect the Panama Canal watershed and tap into local and international capital financing for scaling up the intervention into other communities in the watershed.

Future Water Partnerships to Deliver Services to Urban Poor

USAID's Urban Programs Team is now actively work with other USAID/Washington programs and various Field Missions throughout the world to initiate new CWSF and slum upgrading projects in cooperation with this Facility. For example, USAID is now working with local stakeholders to plan and consider the development of future CWSF projects in Nepal, Ecuador, Jamaica, Mozambique, Angola and Iraq.

3

ENVIRONMENT— SCIENCE AND ENGINEERING

Environment engineering is the technical method used to minimise pollution and other industrial hazards. The permissible limit as approved by Central Pollution Control Board, and State Pollution Control Board of a country is considered by an environmental engineer. It is felt that as we move into twenty-first century, the use of technology problems will increase. Role of engineering and technology in environment matters therefore is now changing in to importance areas.

There areas are (i) sustainable development and (ii) preventive technology. Sustainable development primarily deal with global problems whereas preventive technology is designed to reduce the environmental effects of processes, operations and products. Global sustainable development require a stabilized world population living in secure social and physical environment. These requirements are being badly hit by engineering, technological and scientific practices such as overfertilizing the land, harvesting old grown forests, avoiding pollution control etc.

Preventive technology decisions are guided by market and profit concerns, with little or no consideration for environmental and social at the cost of their health and safety. Environmental science try to explain how diversified pollutions can be checked without causing health hazards to organisms living in this world. So its study in the field of engineering becomes quite important.

Environmental Science and Sociology

Sociology is body of learning about society interacting with social groups. It is a social science which includes the study of social institutions, religions, family, values, customs, caste and habits etc. While living in society, man changes environment very quickly in order to meet his biological and social needs. So on the social front man tries to possess materials for life and for this purpose he removes raw materials from his environment by way of deforestation, extraction of minerals, erection of thermal power stations and construction of concrete jungles for habitat. Indiscriminate construction and expansion of settlement activities collectively pose threat to the environment leading to serious problem of environmental degradation.

Environmental science evolve the study that seeks to describe problems caused by our use of natural world. In addition, it seeks some of remedies for these problems. It tries to protect environment from hit and attach by social- economic activities. Protection of environment from hit and attack by socio- economic activities. Protection of environment with the application of environmental science is as important and focal point before society. It teaches development without polluting the environment with the application of environmental science is an important and focal poind before society. Teach development without polluting the environment. It is agreed that there should be environment development. To this the straight reply is that there should be development within the environment. One should not find these two things separate development and clean. If one cannot breathe pure air, cannot drink clean water and if one has come from west and live on mineral water then there is something fishy somewhere. Birds in this direction assist in communicating environmental pollution because their metabolic rate is much higher than those of humans beings. They predict environmental se up of a region.

Environmental Science and Eco- system

The natural world is organized into interrelated units called ecosystems. Ecosystem is a collection of living and non-living organism connected by a complex web of interactions. Eco-system is a region in which the organism and the physical environment form an interacting unit. This is true in case of lake or island; transition from grassland to desert. Of course it depends on the historical pattern of rain fall in an area. Eco-system look at the ways and means by which natural world is organized. It answers to questions like- where do the rivers flow? What are the typical plants and animals in the area? How does human activity affect the nature?

Environmental scientists recognizes and understand the natural interactions that take place. He integrates these in such a hammer that men make use of this natural world. A few studies made from time to time illustrate this explanation.

North American continent is dominated by intensive agriculture. This means that the original, natural eco-systems have been replaced by managed agricultural enterprise. It is important to mention that this area was at one time wilderness. Today it is difficult to find large regions of true wilderness. The economic value generated by the use of rich soil is tremendous, and consequently most of the land is privately owned. Government cannot control easily what happens on these privately held lands, but do indirectly encourage certain activities. It provides special subsidies to farmers and develop markets for products. Inspite of these facilities, economic risks involved in farming are great. Drought, disease and lack of markets result in failures and on account of this reason, the numbers of farmers constantly decline. Not only this, many communities due to the use of pesticide and fertilizers because they potentially enter the ground water. In addition many farmers use ground water for origination which result in lowering of the water table and less availability of groundwater for other purposes. To stay in this region, to do business and preserve way of life, farmers must make careful use of tools and technology in a responsible manner. Irresponsible use causes increased erosion, water pollution and risks to humans. Environmental sciences teach perfect control and maintenance of eco-system for living.

Environmental Science and Agriculture

There is growing realization that natural farming methods in and answer to the environmental and agricultural problems. The National Academy of Science, Washington has found that farmers who apply few or no chemicals to crops are usually as productive as those who use pesticides and synthetic fertilizers. The group has therefore recommended changing federal subsidy programs that encourage use of chemicals. In the federate corn programme, for instance, farmers are paid a subsidy or roughly $1 for every bushel they produce. The incentive is to produce more bushels, without disturbing environmental condition.

Environmental Science and Economics

Poverty, hunger, malnutrition, unemployment and poor medical facilities are very serious and acute problems in the developing countries. Environmental pollution is another equally serious problem being faced throughout the world due to struggle for economic development. Economic development which require fast industrial development is being made everywhere in the world, in order to needs of teeming millions. To need the needs of increasing population, more and more resources are being tapped from nature. Reckless use of natural resources cause disharmony between human being and the natural environment. This is not economic growth but enactment, hazardous to economic growth. Use of clean technology and pollution free environment ensure higher rates of economic growth because this practice leads to enhanced longevity of life.

A statement issued by the Royals Society of London pointed out that the world population is growing at the unprecedented rate of almost 100 million people every year and human activities are bringing major changes in the global environment. Sustainable development implies a future in which life is improved world- wide through economic development where local environments and biosphere are protected with the application of environmental science assist creates new opportunities for human progress.

Growth and development are fundamentally different goals. Growth results in an increase in material production and consumption that may be unsustainable in the long run : development is a strategy that calls for improvements is culture that do not necessarily require further increase in resources consumption, and environmental degradation.

Environmental Science and Biotechnology

In today's world, biotechnology occupies and Important position in applied sciences in its effort to improve the quality of human life and the living conditions through the development of bio-diversed products, because biotechnologist search nearly all types of herbs through assistance of environmental studies.

It is a observed that the environment come through increased population pressure leading to over exploitation of land, and available water resources, destruction of biowealth, biodiversity a d also injection of various hazardous xenobioti substances into the environment affecting its self-purification potential and carrying capacity. The enormity of the problem has drawn attention of scientist, technologists, administrators and citizens to find ways to sustain and to preserve eco-health along with conservation of invaluable natural resources.

The emerging science of biotechnology is found to be an essential tool in this endeavour it can provide new approach for understanding, managing, perceiving and restoring the quality of environment.

The methodologies of biotechnology not doubt can be suitably utilized to asses the health and well being of ecosystems, transformation of pollutants into harmless substances, to generate biodegradable materials from renewable resources and for developing eco friendly products and manufacturing processes and also disposal methods. Ass these products and process are studied under environmental biotechnology.

Environment Science and Biology

The term biology is mode of two words. One is called bios in Greek which means life and the other logos means study. So biology is the science of living things.

Biology makes to understand ourselves better and the nature in the best possible way. It explain intricate relationship among the organisms in the ecosystem. Studies reveal that about 50,000 children die everyday of various prevented if we commit ourselves to environmental science.

Environmental sanity which means the state of being sane/wise and hygienic. The practising of cleanliness away of maintaining good health which control all the factors in the physical environment. There are a large number of disease which are communicable and non-communicable The various non-communicable disease are hypertension, stroke, cancer, diabetes act. There are many intestirory infectious diseases such as Viral hepatitis, Cholera, Typhoid, Food Poisoning etc. One can protect himself from such disease because health in man's natural condition. It is the result of living in accordance with natural laws pertaining to the body, mind and environment. These laws relate to fresh air, sunlight, diet, exercise. Rest and relaxation of mind. The pure treatments and gifts of environment. The use of these gifts treat mental health and ensure the balanced development of man's personality and emotional attitudes. Worry, low temper, insomnia, fluctuation in mood, cuneus dislike and constant bitterness in feelings. You are always right and others always wrong are a few aches and pains which no doctor can cure to these diseases. So there is interrelationship and interdisciplinary relationship between biology and environment science.

Environmental Decision Making Process

Environmental science teachers that all organisms have an impact on their surroundings. It regnancies development of compromise between different geographic regions. Environmental decisions are therefore often compromising. As already discussed, environment means surroundings. It is a system in which various living beings like animals including man birds, insects, rodents, micro organisms like algae, fungi, protozoa and non-living beings like the soil, the water and the air and interrelated in life sustaining system.

It is a complex web and man occupies a very tiny space in it. Man lives in two environments - one is natural environment of water, air, soils, hills, trees and other living beings etc. The other environment is his social environment. It is an artificial environment, a man-made ecosystem. This environment is created himself alone by using science, technology, religion and politics. Both the environments are lively and lovely, beautiful and bountiful. But some times man meddles too much with Mother Earth to extract much more than what he deserves. Then she becomes crues, hostile and devastating. Environment governs everything that earth provides to people. So environmental problems are people's problems. These problems occur due to unrestricted use of natural resources, which some people feel are justified, result in a diminished environment for other in the region.

For instance insecticides have been applied to crops since they were introduced in the mid 40s to increase yield; prevent pests from devouring food for people more damage to the public health and the environment than the benefits gained from them. There is now a growing body of scientific evidence indicating that pesticides provide only a short term solution to pests and increased agricultural productivity. The two major areas of concern in pesticide use are the crop residues that contain the poison and the phenomenon of resistance that makes insects immune to pesticides. The residues carry not only the portion the kills insects but also the substances that cause cancer, birth defects and many other ailments after contaminating the air, land and water. The tragedy here is that insecticides do not know when to stop killing.

Hot Spot Biodiversity

Studies reveal that wetlands and water bodies in Assam have been identified as a hot spot of biodiversity. It is regretful to notice that wetland in Assam continue to be assaulted by developments. There are many as 3,513 fresh water wetlands. They develop various characteristics on the basis of their location, nutrient dynamic of the soil and vegetation. One of the serious problems associated with wetlands is the construction of embarkments to check floods. The result is the feeder channel which is the umbilical cord of the wetlands is cut off. This action affects the nutrient input and the output ratio between the lake and the main feeder channels.

Consequently, there is an acceleration in the lane use pattern, particularly increased agricultural activities. One of the implications is the change in the fish community composition of the wetlands. A few species have become extinct, leaving this aside, the discharge of effluents for paper mills, woolen factories, refineries and insecticides from agricultural fields have affected the quality of wetland resulting in low productivity and increased incidence of disease both in man and cattle. Scientific management is the only option left for the preservation of Assam's wetlands.

Environmental decision making process occur due to the uses of natural resources. Economic consideration is the backbone of environmental decisions. This is so because a group of people receive value from the natural resources when used whereas another group perceives an economic loss because a use from natural resources been withdrawn. So a conflict is created between two groups. In other words cold war between environmentalist, chemical scientists and economics is generated.

Environmentalists object the use of insecticides and pesticides as well as felling of trees for the construction of river valley projects but economists and chemists do not agree to this proposal. Environmentalists want to protect environment because all these activities affect ecology. The call it mismanagement of resources done in the form of large-scale deforestation, unplanned discharge of residual waste, handling of toxic chemicals, indiscriminate construction and expansion of settlement activities because they collectively pose threat to the environment.

Now the question arises how this dispute should be settled. Environmental decision makers give their views in the following manner.

(i) Some are against these changes; others identify that change must occur.

(ii) Some argue that Government and authorities should take all necessary measures for protection quality of environment, by laying down standards for discharge of environmental degration.

(iii) Some comment that economic considerations should not be given any importance whereas other argue that economic considerations can resolves it.

Now what to do? What potions should be selected what should be. With all these issues in view, compromise is the only way to resolve the conflicts. The social institutions and economic evaluators play an important role in this direction. The field of environmental science seeks to find the middle path and ground. The environmental science teachers the impact on environment when natural resources are converted into goods. It is therefore, argued and suggested that before natural resources are converted into artifacts that is when deforestation is exercised for making diversified use of wood and elephants are killed to get ivory, what impacts they have upon environment should not be forgotten. Environmental impacts while underrating River Valley Projects in no way less important because this action lead to degradation of catchment areas, rehabilitation of those who are affected by environmental changes, increase in water borne disease, reservoir inducing seismicity and causing loss of flora and fauna. To minimize these changes some standards are framed. International standard organization has framed some standards for better environmental management. India has also adopted IO 14000 which incorporate series of management standards. The components of ISO 14000 are as follow:

a. Environmental Management System :- It is a system for maintaining development in the environment. It's aims are recognizing impacts and risks of environmental degradation and developing all environmental control policies so that environmental performance is improved.

b. Environmental Performance Evaluation :- ISO 14000 is applicable to all types and sizes of organization. Its applicability is traceable at irrigation Department, Forest, Extraction of Oil and Natural Gas and various industrial units. Any organization where nature of activity and service have direct and indirect effect on environment can adopt these standards. The success of this system depends on commitment from all level. EPE unit try to achieve standards as per ISO standards after examination an d evaluation.

c. Environmental Auditing :- It aim is maintaining programmes and procedures for periodic EMS audit. Auditing is done to know whether or not the environmental management system conform to the Environmental management standards.

d. Environmental Aspects of Products Standards :- It includes element of an organization' activities, product or services which can interact with environment.

e. Life- cycle Assessment :- Under this the organization establishes and maintains documented procedures to monitor key characteristics of its operations and activities that can have significant impact on the environment.

e- Environmental Labelling :- It ensures conformance with stated environmental policy and make a self determining declaration of conformance with the standard.

Environmental Impact Assessment

Environmental impact means effect on environment due to human activities directed in pursuit of fulfilling ever increasing needs of life. Assessment refer to evaluation and estimation of exploitation and processing of natural resources. The EIA is therefore a method of evaluating environmental consequences i.e. environmental changes which are likely to be caused by the proposed human activity related to land use construction of dams, reservoirs, industrial location, urban expansion etc. and the possible adverse effects of these environmental changes which may cause ecological imbalance and ecosystem dis-equilibrium etc.

Methods of Checking Environmental Impacts

(i) Polluters should pay for their environmental impacts through internationalization of costs, the use of appropriate taxes, changes and incentives for environmental protection.

(ii) Unsustainable patterns of production and consumption must be reduced and appropriate demographic policies must be promoted.

(iii) Environmental protection must be an integral part of development.

(iv) Creation of an agency to prepare an environmental impact statements for analysing the major effects of projects to be undertaken. EIS will reconsider a harmful project.

(v) Society need urgently many more teacher at every level who should be trained in environmental education. Outdoor activities and natural sciences are important components of this mission.

(vi) Outdoor activities are an essential part of environmental education, which strives to introduce environmental literacy to the general population and to encourage students to prepare for environmental careers.

(vii) Providing environmental literacy - a sense of duty to care for and managing wisely our natural endowment and productive resources for the long haul.

(viii) Encouraging pose secondary students to pursue careers related to the environment, appreciation of our natural environment, knowledge of basic ecological concepts, acquaintance with broad range of current environmental problems.

(ix) Improving understanding among the general public of the natural and built environment.

(x) Promoting efforts to prevent damage to the environment and the biosphere.

(xii) Environmental careers are required to be expanded because environmental studies is a very vast subject. So far perfect assessment and evaluation environmental professionals such as engineers are needed to develop technologies and products to clean up pollution, economists , geographers, and social scientists are needed to evaluate and other specialists to develop government and industrial policy, laws, and regulations to protect the environment. The EPA estimates, for example, the 100,000 new professionals will be required in the United State in the next five years alone to deal with hazardous waste problem.

Dharma Ecology

On the occasion of the World Environment Day, one aminate environmentalist presented he concept of Dharma on Ecology. According to him there is a need to have global ethic to meet threat to our environment. Planners must make use to this concept and educate the people over this ground, because the major challenge facing the human race is to move towards susceptibility.

Following six principle of Dharma Ecology as recognize by environmentalist are mentioned below:

(i) Halting and revising the arms built-up.

(ii) Meeting genuine social needs.

(iii) Willingly adopting sustainable way of life.

(iv) Crating awareness about concealed social, economic and environmental costs.

(v) Protections and nurturing renewable resources, conserving non- renewable resources and avoiding conserving non-renewable resources and avoiding waste by exercising economy.

Need for Public Awareness

The enhanced pace of developmental activities and rapid urbanization have resulted in stress on natural resources and quality of life. The trend of increasing pollution in various environmental media is evident from the deteriorating air and water quality, higher noise levels, increasing vehicular emission etc. Realising the urgent need for arresting the trend, Ministry adopted Policy for Abatement of Environmental Degradation. But this is only a approach and not a movement. Till movement is not effect, approach shall stand null and void. To make movement effective public support is very necessary. So a clarion call of public awareness becomes essential. There is no urgent need for public awareness about cleaner environment.

It is, therefore suggested that while imparting environmental education special attention must be paid to school going children and women because they alone constitute about 50% of the population. The should be made aware of heath, nutrition, sanitation, hygiene water and food contamination. The main objectives of environmental education should be (i) improving the quality of environment (ii) creating an awareness among people on environmental problems and conservation of (iii) creating such as atmosphere so that people find themselves fit enough to participate in decision making so that developmental programmes are presented after evaluation.

However, a number of environmental management

approaches have bee developed with the lapse of time and the same are discussed below in a lucid manner.

Scope and Importance of Environmental Studies

Scope means coverage, end aimed at, purpose, intention, outlook, purview and sphere of observation. The scope of environmental subdues is vast and wide. Sky is its limit and ocean is its depth. What is not covered in environmental studies shall take years and years yet to think as how to cover. Studies reveal that during the past, few decades there has been fundamental change in the attitude of man towards environment. There was time when environment meant only sanitation and health.

Today, the environment is conceived in its totality and a holistic approach is designed while planning a better quality of life stressing upon sustainable development. Who with not agree with this fact that human activities have been changing with greater speed and velocity. Our number population- wise is changing fast.

Environmental studies Teachers us deforestation and forest conservation

Increasing population, along with widespread poverty, has generated pressure on our natural resources. Of the 329 m ha. of the total land area in the country, it is estimated that about 174 m. ha. is degraded; this consists of agricultural as well a non- agricultural lands and forest resource. The forest resources are threatened due to overgrazing and other forms of exploitation, both for household and commercial needs, encroachments, unsustainable practices like unscientific cultivation and degradation of forests and tree lands. Environmental science teaches how forest can be covered. National afforestation and Eco-development programmes provide methods of forest wealth conservation, because trees and plants are the life and blood of human civilization. They have served and are serving man-king in various ways from days of yore Media Parkar Sunder Bahuguna, the leader of the Chipko Movement, rightly pointed out that preservation of trees and plants alone can save the human species from environmental pollution. The fist and foremost advantage of forests is that they absorb carbon dioxide, provide water and control soil erosion. To save trees in to save civilization.

(ii) Environmental Studies Teaches us Diversified Ecologies

Conservation ecology deal with the application of ecological principles to the proper management of resources leading to sustained yields of useful resources to human welfare. Resource ecology deals with the renewable and non-renewable resources and their judicious management. Pollution ecology deals with problems associated with the movements of pollutants in the environment, environmental deterioration and the maintenance of its cleanliness. System ecology deals with the analysis and modeling of ecological systems. Radiation environment. Paleoecology deals with organisms and their environment in the geological past.

(iii) Environmental Studies deal with air pollution check-up - Chemical and allied industries are the major source of air pollution because these industries discharge tons of noxious gases. With the increasing standard of living and a growing end for a faster moving life, the numbers of automobiles on the Indian roads have been increasing and simultaneously so is the pollution from them. The exhaust of automobiles and diesel driven heavy vehicles contain many harmful gases and the problem of air pollution is more significant at found level. Since the major activities of human beings are at this level, the pollution from automobiles has been gaining notoriety day by day. It has been determined that they contribute 39 percent of the total air pollutants in terms of tonned is the USA. Air pollution caused by motor vehicles contaminate surroundings. Smoke, odour and irritating compounds are causes of nuisance, especially to a person standing next an idling vehicle. The cumulative effect of many vehicles, plying especially on busy roads during peak hours has highly deteriorate effects. Those residing in the vicinity of such roads are worst affected and they often suffer from chronic heart and oxides of nitrogen, and under certain favourable conditions of temperature and sunlight, certain photochemical reactions are highly toxic, irritate the eyes and the respiratory system, damage vegetation and impair visibility. Today the number of planes, rails and vehicles which emit carbon mono-oxide along with smoke is much larger than earlier. According to one estimate the oxygen consumed by one automobile after covering a distance of 960 km. is enough for a man for one year. Tokyo is the most polluted town in the world.

Every policeman there is provided with an oxygen cylinder so that he may use it when the need arises. In Los Angeles people are advice not to undertake any exercise and inhale air in foggy environment. The toxic vehicular exhaust are sources of considerable air pollution. The ever increasing vehicular traffic density pose continuous threat to the ambient air quality. It is estimated that in all the major cities of a country about 800-100 tonnes of pollutants are being emitted into the air daily; of which 50% comes from automobile exhausts alone. The sources of emission in atomiles are: (i) Crankcase , (ii) Fuels tank and carburetor.

Broadly two types of air pollutants- primary and secondary air pollutants are emitted. Primary air pollutants are carbon monoxide, hydrocarbons, particulates, sulphur dioxide and nitrogen compounds. These five pollutants are released into the atmosphere in an unmodified as well as in sufficient quantities which pose hazards and risks to human health.

Environmental science teaches how to maintain AAQ and keep air pollutants within the permissible limits. Replacements of diesel driven heavy vehicles with CNG is one important attempt in this direction. Hats up to environmentalists.

(iii) Environmental studies is useful in controlling water pollution - In India, there are several river systems which accounted for 83% of the land mass, 80% of the population and 85% of the water discharge. A survey report reveals that 90% of pollution load in the river systems in on account of human wastes. The different types off diarrhoea, which in themselves may not be killers, but could prove very dangerous when accompanied by malnutrition and about 3 mullion children die per minute. Water borne disease occur primarily due to the lack of proper environmental sanitation.

We are told that almost 13,000 litres of blood enter everyday in Yamuna because of slaughters. More than 100 different virus types are known to be found in human refuse. Viruses are quite common in sewage. They can survive for months in water and soil. Recently a number of diseases have been linked to contaminated water are increasing. The contaminated water use for irrigation purposes has adverse effects on the fruits, vegetables that we consume in our day to day life. The extension of irrigation is creating excellent breeding grounds for disease by leaps and bounds. Intensive agriculture based on chemical agents for fertilizers and crop protection has also become a major cause of water contamination by ammonia.

The major causes of water pollution are sewage and other waters, industrial effluents and agriculture discharges. Broadly biodegradable and non-biodegradable pollutants are responsible for water pollution. Environmental engineers suggest the various ways of controlling water pollution as follow:

a. Sodium Salts : These could be removed by reverse osmosis method. Sodium sulphate from a rayon mill effluent cab be easily removed. The water for reuse cab be recovered by this method. From Cheese whey, we may recover 70% of protein a d 80% of lactose by this method.

b. Decolorization of water : The waste water from printing and sari dying industries could be decolorized by an electrolyte decomposition technique.

c. Phenolics : This can be removed from waste water of pulp and paper mills, carbonization plants, petroleum refineries, tanneries and resin plants by the use of polymeric absorbents.

d. Mercury : This could be removed from Chloro-alkali effluent plants by using mercury selective ion-exchange resin.

e. Ammonia : This cab be removed from waste water of industry by ion- exchange technique.

f. Re-utilization and recycling of waste : Different types of wastes such as industrial effluents, sewage and sullage of municipal and other systems and thermal pollutants can be recycled for beneficial use. Urban waste may be recycled to generate cheaper fuel gas and electricity.

Technology has been developed for magnesium not of radio-active wastes and chemical wastes. In Okhla, New Delhi is large treatment plant for sewage recycle is already in operation. Waste water reclamation through aqua culture, utilization of domestic and industrial waste water in agriculture and detoxification of phenol is a suitable technique developed. On distillery in Gujarat is able to treat 4,50.000 litres of waste daily and generate energy equal to that produced in 10 tonnes of coal.

g. Stabilization of the ecosystem : It is the most scientific way of controlling pollution and is done by reducing waste input, harvesting, removal of biomass, trapping of nutrients, fish management not and aeration. This can be done by biological and physical methods to maintain ecological balance.

Environmental studies is very useful for ecological health

Environmentalists teach disposal of the dead and burning. Provision of gas or electric crematoria have been made in all major town and cities. Planting of trees in an integral part of school and college education. No student should be given his degree or diploma unless he or she produces a certificate of having planted and nurtured at least ten or twenty trees during his or her students life.

We must understand the nature of human cells and the nature of plants cells which provide nourishment and around which animals and man have evolved. It is the quality of plant life on which the quality of human life depends. Ecological health depends upon keeping the surface of earth rich in humus and minerals so that it can provide nourishment and around which animals and man have evolved. It is the quality of plant life on which the equality of human life depends. Ecological health depends upon keeping the surface of earth rich in humus and minerals so that it can provide a foundation of health property and cultural stability. When the trees are in plenty there is a natural cycle involving the trees and the soil. In autumn the trees drop their leaves returning them to the earth where they decompose. A complex mixture of bacteria and fungi inhabit the soil. These microbes plan an extremely important role in the biochemistry of the soil.

It is a well known fact that after the ruthless deforestation, it is impossible for us to have plant food grown without the help of artificial fertilizers for the millions of people in our country and suffering from the consequences thereof. The most important measures of population control is the grow more and more trees in the hills and foothills, plains and in clusters in the fields everywhere and every nook and corner of the country on a war-footing. The symbolic rituals of getting a few trees planted by visiting dignitaries here and there, is of no use.

4

BIO-DIVERSITY AND ITS CONSERVATION

Introduction : Definition of Bio-diversity

The term biodiversity is made of two words- bio and diversity. Bio means living things whereas diversity means state of being varied or variety. So in nutshell biodiversity means varied type and king of species. Different organisms and species are examples of bio-diversity. Earth's bio-diversity is organized into three levels :

(i) At the top are the ecosystems such as rain forests, coral reefs and lakers.

(ii) At the bottom are the variety of genes, making up the heredity of each species.

(iii) Third down are the species that compose the ecosystems such as swallow tail butterflies, moray eels people.

Now one question arise in the minds of readers as how much biodiversity is there. Biologists have described a total of between 1-5 million and 1-8 million species. Yet this impressive achievement is only a small beginning. Estimates of the true number of living species range, according the method employed, from 3-6 million to more than 100 million. Least known are the smallest organisms. By repeated sampling, biologists estimate that as few as 10% of the different kinds of insects, worms and fungi have been discovered for bacteria and other micro-organisms, the number could be well below 1%. Even the largest and most intensively organisms are incompletely catalogued. Four species of mammals, for example have recently been discovered in the remote Aminate Mountains along the Vietnam- Loas border. One of them, the saola or spindlehorn is a large cowlike animal distinct to be classified. Earth, as far as life is concerned, is little known planet.

Genetics

Genetics is the study of the laws of inheritance. The first scientific study on genetics was carried out by Gregor Mendel an Austrian Monk, biologists and a famous geneticist. Being son of a successful horticulturist, Mendel developed a love for gardening. His father initiated him into the art of grafting of fruit and trees. Later he entered a monastery in Bruin, Where he carried on his scientific investigations on hybridization of plants. Mendel died before he could see his work accepted and appreciated. Genetics is the modern experimental study based on the discovery of Mendel's experiments.

Genetic is the study of heredity and variation in animals and plants. Every organisms carries genes. This is true of bacteria, fungi, plants, animals and human beings. These genes are called codes. These codes are nothing but massages. They carry information about organisms as what chemicals are needed to grow, survive and reproduce. Genetics try to solve puzzle of what genes are, how they are duplicated and transferred. Genetic engineering studies about moving of genes form one organism to another. It is the manipulation of heredity material. It seeks to change not only the species that provide us with food and drink but also those involved in sewage disposal, pollution control and drug production. It also seeks to create micro-organisms, plants and animals that can make human products such as insuling and even possibly to change the genetic make-up of humans.

Contribution/Applications :

The genetic engineering hopes to provide society with a vast range of innovations such as-

(i) Biologists use genetics to identify the genenes that function in the life of the cell and those control the development of complex organism from a fertilized egg.

(ii) Rapidly growing biotechnology industry is using genetics to produce a range of products from pharmaceuticals to micro-chips.

(iii) In agriculture, genetics is the basis of breeding new crop plants and livestock.

(iv) Scientific investigations seeking to understand the molecular basis of genetic diseases and to effect its cure.

(v) Providing counselling to families who are at risk of producing children with genetic defects.

(vi) Understanding the inheritance of genetic diseases.

(vii) Diagnosis and correction of genetic disorders.

(viii) New approaches to the generation of energy.

(ix) Recovery of resources and pollution control.

(x) More abundant food products.

(xi) More effective and cheaper pharmaceutical products.

Species

Species literally means kind. Species are all the organisms of the same kind able to breed in nature and produce live fertile offspring. Species also include plants- the definition of which provide exception but for our purposes this is a useful working definition. Species living in a given area at the same time consists of a population. Problems are faced with species definitions based on reproductive isolation. Investigators mostly cannot know whether two groups that live in different places are capable of interbreeding. Species identification therefore becomes a serious problem. Morphological characteristic such as size, shape, colour and skeletal structure is the only way opened for species identification.

How many species are there on Earth is an important question. Though it is difficult to answer yet some scientists confidently declared that every important king of living thing on Earth would soon be found and named. Most of these findings concentrate on species such as birds and mammals. Recent studies of organisms such as insects and fungi suggest that millions of such species and varieties to be studied scientifically. Based on new discoveries and research taxonomists estimate that there may be somewhere between 3 million and 50 million different species alive today.

In fact, there may be 30 million species of tropical insects alone. About 70 percent of all known species are invertebrates. This group perhaps makes up the vast majority of organism yet to be discovered. They may constitute 95 percent of all species. We now believe that the 2-1 million species presently known represent only a small fraction of the total number that exist.

Plant Diversity

Animals and human beings are deeply indebted to plants on account of their following diversified uses :

(i) Green plants, some of them when appear in the shape of forests covering 22.7% of the total land area of

India provide natural habitat and food to wild life, raw material for use in industries; supplying oxygen gas during photosynthesis; checking soil erosion by way of binding re-roots; cool atmosphere and influence the local rain-fall.

(ii) One has rightly asserted that the existence of man would have been unthinkable and impossible without plants as they provide oxygen for respiration and food for nutrition.

(iii) Plant exercise monopoly by way of preparing food such as starch a chief constituent of rice, wheat and potato etc., by obtaining carbon dioxide from the air, water and inorganic salts from the soil.

(iv) Green plants purify the atmosphere by absorbing carbon-dioxide from it and releasing an almost equal amount of pure oxygen into it.

Biodiversity in India

India's biological diversity is quite immense. She has over 45,000 plant species which represent about 7% of world's flora and 6.5% of world's fauna. India is one of the 12 countries identified as mega-centres of biological diversity.

Embracing 10 biogeographic Zones and 25 biotic provinces, India incorporates an exceptional range of bio-diversity, including all the world's major ecosystem. As one of the eldest and largest agricultural societies, India also has an impressive diversity of at least 166 species of crop plants and 320 species of wild relatives of cultivate of cultivated crops.

(i) Wild Life - Regarding wild life presently, the protected area network comprises 84 National Parks and 447 wildlife sanctuaries. The Protected area covered is 4.5% of the total geographical area of the country. There are over 200 Zoos and 23 Tiger reserves in the country.

Financial, technical and scientific assistance are provided to states. The national Wildlife Action Plan adopted in 1983 provides the frame work of strategy as well as programme for conservation of wildlife.

10 to 15 percent of all the world's species live in North America and Europe. New species of higher plant and animal are rarely found in developed countries of the world. Tropical areas are the centres of greatest biodiversity. They have never been studied by scientists thoroughly. The Malaysian Peninsula, for example has at the least 8000 species of flowering plants, Britain double the area has 1400 species. South America has fewer botanists available to study perhaps 200,000 species of plants. Twelve countries, mainly in the tropics, account for at least 60-70 percent of the world's bio-diversity.

(ii) Coral reefs - Coral reefs are shallow water tropical marine ecosystems characterized by high biomass production and rich flora and fauna diversity. The Four Indian coral reefs are : Gulf of Manner, Andaman and Nicobar Islands, Lakshadweep Islands and Gulf of Kutch. The survival of resources is threatened by factors like oil spillage etc., and often because they fail in the shipping route.

(iii) Fauna - Mammal fauna comprises 372 species, with 63% found in Assam, India's 1,228 bird species represent about 13% of the world's total. Reptilian and amphibian fauna include 446 and 204 Species respectively, with concentration in Western Ghats. Faunal endem's varies substantially from 3.5% in mammals and 14% in birds to 32% in reptiles and 62% in amphibians. Among the larger animals, 101 of birds. 15 of reptiles, 3 of amphibians and 2 of fish are considered endangered.

(iv) Flora - Indian flora comprise about 15,000 flowering plants, an estimated 33% of which are endemic and roughly 1500 plant species are threatened. Areas rich in endemism are the North-East, the Western Ghats and the North-Western and Eastern Himalayas, however, the Andaman and Nicobar island contribute at least 200 endemic Species.

Universal Ethical Principles

There are four schools of thought which respond to moral laws such as Universalists, Relativists, Nihilists and Utilitarians. The thinkings of these schools towards environment are explained below turn in a nutshell manner.

(i) Utilitarians - Bentham is called utilitarian. He was English philosopher. Bentham was an eccentric genius. He equated goodness with happiness, and happiness with pleasure. Studies reveal that the views of Bentham. He regarded pleasure as the only thing worth having in its own right.

(ii) Nihilists - Thinker Schopenhauer is recognized Nihilist. In spite of living in a civilized and comfortable society. Schopenhauer painted dark picture of the world. He commented that everything is arbitrary and there is no purpose in life other than the dark and unceasing struggle for existence. So according to the views of Nihilists there is no reason to behave morally. We live in the world of uncertainty, pain and despair and there is no such thing as good life, Might is right eat or be eaten.

(iii) Relativists - Sophists are called relativists. The oppose views of Plato. The opine that moral principles are relative to circumstances and are always contextual. They relate to a particular person, society, or situation.

(iv) Universalists - Plato and Kant talk about the universality of ethical principles. According to these moral philosophers fundamental principles of ethics are universal and eternal. These rules of right and wrong are valid regardless of out interests and preferences. Some believe these rules are revealed by God while others maintain that they are discovered through reason and knowledge.

Social, Ethical, Aesthetic values

Ethics is one branch of philosophy. It seeks to explain the difference between right and wrong, regardless to cultural attitudes. Now one feeling must be creeping in out minds as what ethics has to do with environment. To this is replied that there is definite interrelationship between ethics and environment. All environmental issues involve consideration of ethics. Sufficient food and fabric is available in the world to feed people. It is unethical to allow some people to starve and others consume more than what they need.

Developed nations do not realize it. They have enough and still they need more. Their greed for more is ever increasing. They do not surrender before morals and ethics of life and share with others what they have. The must think less about conquering nature and more about learning to work with nature. Each person must realize his interdependence with the rest of he nature including his fellow human beings. To safeguard life on earth, people must learn to control and adjust the balances in nature that are altered by their activities. One more reason is placed in this context. Members of the society very well realize this view. They view the world's energy situation as serious and have reduced their own consumption. But may do not believe and realize this problem. They continue to consume energy as long as it is available and do not accept any cut on the use of it. Other similar issues are the population problem and pollution. It is ethical to have more than two children when the world is facing problem of over-population. Similarly an industrialist may not probably look on the negative aspects of pollution compared to one who is outdoor. Some sixty American cities have been declared having unsatisfactory water supply due to toxicity of pollutants released by industries.

Every year in the United States, 20 million tones of paper and 200 million tons of domestic garbage in junked. Keeping this in view the Massachusetts Institute of Technology, USA reveals that there shall be disastrous collapse in the coming hundred years if both population growth and industrial growth are not brought to an immediate halt. So the first task is to educate economic harpers about the value of environmental ethics. Clean and unpolluted air away from the chimneys of industries, free from fumes and smoke has a great cleaning and purifying effect for sustainable growth. For this purpose, taking ethical stand is important regardless to those who disagree with it.

Emerson, a pioneer philosopher environmentalist in his first essay Nature published in 1836 objected fast economic development by way of invading nature. Another naturalist who held the beliefs similar to Emerson was Henry David Thoreau. According to Thoreau, Truth in nature and wilderness over the deceits of urban civilization. He summarized his feelings about nature as But most men, it seems to me, do not care for nature and would sell their share in all her beauty, as long as they may live, for a stated sum-many for a glass of rum. He further concluded Thank God, men cannot as yet fly, and lay waste in the sky.

Rachel Carson rank above all naturalists of her time who generated widespread discussion about pesticides and its potential dangers to food, wild life, and humans in her work Silent Spring published in 1962. Carson's early death from cancer came before her book was recognized, establishing pesticides can contaminate and cause widespread damage to the ecosystem. In style and content, Rachel Carson designed Silent Spring to shock the public into action against the misuse of chemical pesticides. She described the poisons, pointed out the failure to grasp biological principles that allowed us to direct broadsides of these poisons against the environment and detailed the resulting disasters. According to one report DDT tends to accumulate in the fatty tissues of wildlife. In 1957, the waters of California's Clear Lake were found to contain only .02 parts per million of DDD, a close relative of DDT. Microscopic plants and animals in the water stored residues at five parts per million. Yet fish eating large quantities of microscopic organisms, concentrated these residues to over 2,000 parts per million. Grebes which fed on these fish died in great numbers.

Ethics being a part of philosophy relates to morals and values. Morals tell difference between right and wrong whereas values are concerned with the ultimate worth of actions or things. Western countries consider the roots of morals and values as posed by Socrates and the Greek philosophers 2500 years ago. According to these philosophers what is the good life and how should we as moral beings behave. Environmental ethics asks about the moral relationships between humans and the world around us. We know that besides human beings, we are surrounded by species and nature. Do we have ethical principles, duties and responsibilities towards these. Are these ethical principles that constrain to use resources or modify our environment. Are we not going against the nature? Do we listen what nature speak to us? Nature will not speak, nature goes its way, headlines and heartless. No matter whether man responds to it or not. How are our obligations and responsibilities to nature weighed against human values and interest? Do some interests or values supercede others? These are very important questions posed to us.

Values, Rights and Obligations

Human beings are moral agents, because they are capable of acting morally or immorally. Capabilities that enable humans to form moral judgements include moral deliberation. Moral deliberation are capacity to carry out decisions and the responsibility to answer all those right things which he fails to carry out. All humans do not have all these capacities, children, mentally retarded and mentally ill are not regarded moral agents. If a child murder someone, we do not hold him responsible, because children are considered moral subjects.

Weaker members of society dissever equal treatment with those who are stronger was not universally accepted. Powerful people always considered women, children, outsiders and serfs as their property. Even Greek philosophers accepted slavery as natural and necessary. But on ethical grounds it is incorrect. No one can ethically treat another human as a mere object for their own pleasure or profit.

Like human beings, do other species have some rights is an important question covered in environmental ethics. Though non-humans are encompassed in moral extensionism yet some philosophers do not agree and claim animals automa. Pet owners claim of course while animals may not have clear feelings. This is all about organisms but what about bugs, rocks and landscape. Should we extend moral considerations to these non-sentiment beings is an equally important with regard to ethics. Value measure the worth of something. All human beings have inherent value because they are human. Tools on the other hand have instrumental value. Tools have value because those who use them have value.

Now the question arises as do non-sentiment things have inherent value. Non-sentiment things are non-humans. Some people believe that even non-living things also have inherent value. Rocks, rivers, mountains, landscapes and earth certainly have value. These things exist since time immemorial even before our arrival and we cannot recreate them if destroyed . In a landmark, 1969 court case, the Sierra Club sued the Disney Corporation on behalf of the trees, rocks and wildlife of Mineral King Valley in the Sierra Nevada Mountains where Disney wanted to build a Ski resort and the Ski was never built. Expansionists and ecological world view give the following statement in support of non-sentiment thins :

(i) Development means essentially the same thing as growth.

(ii) Technological innovations can sustain an ever-increasing human populations.

(iii) Progress means the satisfaction of increasing levels of human wants.

(iv) Nature is a storehouse of raw materials to be used to satisfy ever increasing human needs.

(v) Earth's resources can support unlimited economic growth.

In view of above statements many tribals both hunters

and gathers have a strong sense of responsibility to act as custodians of resources. They feel to have proper role as working together with human and non-human forces to sustain life. Human are seen as partners in natural process rather than masters - not outside of nature but part of it. Egocentric claims moral values and rights for ecological processes and systems. According to them if you kill an individual organism, you deny it a few months or years of life, but if you eliminate and entire species or a whole landscape, you have destroyed something that took millions of years to create.

Value of Bio-diversity

We benefit from biodiversity in many ways. Some of them are discussed below :

(i) Drugs and Medicines : Living organisms provide useful drugs and medicines. More than half of all prescriptions contain some natural products. The value of pharmaceutical products from the plants and animals of third world countries are accounted to be worth $30 billion according to the UNDP. Indigenous communities that have protected and nurtured these organisms on which these products happens to be based are much less compensated. Realize the success story of vinblastine and vincristine. They are anticancer alkaloids. They are derived from the rosy periwinkle from Madagascar. This organism provide anticancer drugs. It use inhibit the growth of cancer cells and is found very effective in controlling cancer. This drug also make childhood leukemias and Hodgkin's disease highly remissible.

Pharmaceutical companies are actively engaged in search of herbs in tropical countries. Merck-world's largest biomedical company is paying $1 million to INBIO of Costa Rica for supply of plants. INBIO also impart training to natives to locate and catalogue all the native flora and fauna between 500,000 and 1 million species in Costa Rica. This is good scientific information besides sharing profits from native resources.

(ii) Human Food Supply - Human Food of everyday use comes from other organisms. So many wild plant species therefore make important contributions to human food supplies. According to noted ecologist Norman Meyears nearly eighty thousand edible wild plant species for food and medicine beings. Some four thousand native plant, animal species for food and medicine are used by villagers in Indonesia. Indonesia has 250 edible fruits and only forty-three of which are cultivated widely Academy of Science. Studies reveal that unfortunately human disturbance are destroying valuable food supplies and domestic crops before identification and prevention of their genes on account of the following :

1. Conversion of natural landscapes to agriculture.
2. Forest clearing.
3. Overgrazing.

(iii) Aesthetic and Cultural Benefits - The diversity of life on this planet open many doors of aesthetic and cultural benefits to man in innumerable manner. Millions of people enjoy hunting, fishing, hiking, wildlife watching and other outdoor activities based on nature. These activities provide invigorating physical exercise. Observing and protecting nature has religious or moral significance for many people. Nature's appreciation is economically important also. Note the important information communicated by U.S. Fish and Wildlife Service :

a. To enjoy nature and experience other cultures inexpensive intercontinental air-fights are arranged to visit remote and exotic planes in USA.

b. 8 million on cinema tickets and $ 5.9 billion per year compared to $ 5.8 billion on cinema ticket and $ 5.9 on sports events.

c. Americans spend $ 18 billion every year watching wildlife.

(iv) Ecological Benefits - Biodiversity provide ecosystem stability. Soil formation, air and water purification, nutrient cycling, solar energy absorption and management of hydrological cycles all depend on the biodiversity of life. Nature as a matter of fact maintain ecological processes automatically with no cost to us. Besides this it provides a genetic library of information about ecology which man could never reproduce.

There have been a great deal of controversy about the role of biodiversity in ecosystem. Ecologist like David Tillman opine that plant communities withstand environmental stress better and recover more quickly than those with fewer species. But this is not agreed in toto. Since we do not understand fully the complex interrelationships between organisms so we are often amazed at the effects of removing insignificant members of biological communities. Wild species provide significant services in suppressing pests and disease carrying organisms. It is estimated that 95% of potential pests and disease-carrying organisms in the world are controlled by species.

The PRAYING MANTIS is an effective predator against garden pests and is harmless to human. Ladybird beetles prey on a variety of pests both as larvae and adults. For a few dollars you can buy several thousand of these hardy and colourful little garden protectors from gardening supply stores. So by preserving natural areas and conserving wild species, we utilize the stabilizing diversity of nature that keep pests organisms in balance. Hence Bio-diversity is beneficial in preserving ecological balance.

Biodiversity Threatens

Extinction of species is of great concern because despite our increased ability to alter and restructure our surroundings, people depend on other species is many ways. Extinction eliminates species from the planet and this elimination is a natural process. They eliminate in many ways such as species die out and are replaced by others, i.e. by their descendants. Thousands of species disappear due to human interference because man disturb ecosystem. Not only species but sub-species also extinct every now and then. Though it is difficult to provide accurate data on species distinction yet one feeling is very clear that if present trend continue, we may destroy millions of kinds of plants, animals, and microbes in the next few decades.

(i) There are two distinct forces which express concern about the conservation of Bio-diversity. One is recognized as ecological conservation and the other commercialized conservation. Ecological conservation of biological diversity arises due to severe genetic erosion. Genetic erosion takes place as a result of commercialization of agriculture and forestry. Genetic erosion is an ecological hazard because it leads to the extinction of life forms which have value in themselves.

(ii) Endangered creatures called Olive Redley Turtles are being slashed by the propeller blades of fishing trawlers or caught in fishing nets. The death toil of the turtles from 1991 was 50,000.

(iii) 70 percent of all recorded wolf mortalities in Kutch are due to spraying of position around their dens.

If this trend continues, we will soon lose the entire carnivores guild of Kutch. Wolves, even although they are known to be hardy race, are on the endangered list of wildlife. Earth is one intricate ecosystem of links by which all life is shaped. Lose one species, and a thousand others will be on the brink eventually threatening our survival.

Human caused reduction in bio-diversity

The rate of species's extinction is dramatic for the last 150 years. Between A.D. 1600 and 1850, disparaging of species from this planet has been due to alarming increase in human activities. According to Harvard university entomologist E.O. Wilson, we are pushing 20,000 or even more species per year into extinction. This is only a tentative data about the extinction of biodiversity because many parts of the world are not thoroughly explored and many species might have disappeared before they were studied and classified by biologists. The greatest loss of extinction is to tropical zones of the world because 90 percent species belong to this group. Human caused reduction in biodiversity is exercised by way of habitat destruction, hunting, fishing, commercial products and live specimens. The details of these reasons are given below to facilitate understanding of readers :

a. Genetic Assimilation - Endangered species like mammals and birds crossbreed with closely related species and opportunistic plants introduced into a habitat or displaced from their normal ranges by human actions genetically overwhelm local populations. This process is called genetic assimilation. Many endangered species are threatened by genetic assimilation because they crossbreed with more vigorous related species. Hatchery raised trout are introduced to streams where they genetically dilute indigenous stocks. On account of this reason black-ducks have declined severely in USA and Canada.

(ii) Pollution - Toxic pollutants have disastrous effects on local populations of organisms. Declining marine mammals, alligators and fish populations show complex interrelations between pollution and health. Widespread deaths of seals on both sides of the Atlantic are due to persistent accumulated chlorinated hydrocarbons such as DDT, PCBs and toxins in their body which weekend their immunity system. Similarly deaths of pacific seal loins and striped dolphins in the Mediterranean are thought to be caused by accumulation of toxic wildlife. The U.S. Fish and Wildlife Service estimates that 3000 metric tons of lead-shot are deposited annually in wetlands and that between 2 and 3 million waterfowl die each year from lead poisoning. Scavengers such as bald eagles eat birds and mammals that have lead shot or bullet fragments in their bodies. Though lead-shot is banned yet spent shot remain in bottom sediments of lakes, ocean and water-reservoirs where bottom feeders consume it.

(iii) Diseases - Pathogens and disease organisms are also recognized as predators. To be successful over a long period of time, a pathogen must established a balance in which it is vigorous enough to reproduce, but not so lethal that it absolutely and completely destroys its host. When a disease is introduced in the new environment, this balance may be lacking and an epidemic may seep through the area.

The American chestnut was one very popular amongst hardwood forests. One among four trees was chestnut. This tree was over 45 m tall and 3 m in diameter. Its nutritious nuts were important for birds particular for passenger pigeon, forest mammals and humans. The wood of this tree was quite useful for making fence posts to fine furniture. In 1904s, a shipment of nursery stock from China brought a fungal blight and within forty years, the American chestnut disappeared. A similar disaster has devastated American elms.

(iv) Induction of Exotic Species - Exotic species are non-native organisms. They are aliens. They are both animals and birds. They have been introduced intentionally and accidentally. Zebra mussels and spiny water flues are introduced accidentally. English sparrows and common pigeons are aliens and introduced by acclimatizers who spent considerable sums of money to import these species from Europe. Exotic species in the United States. Alien species from around the world now threaten native species throughout the Great Lakes. Probably the most endangered group in the United States are fresh water mussels, of which more than half of the 297 known species are threatened, endangered, seriously declining or already extinct. Pollution, siltation, and competition from exotics like zebra mussels are largely responsible. Some of the exotic organisms that have, with human help invaded the Great Lakes. These aliens are driving out native species and irreparably damaging the ecosystems.

Plants when introduced into a new environment also multiply explosively. Kidzu-a legume that grow in Asia when brought in United States grow and run with so much of wild expansion that it covers trees, houses and whole hillside. Not only that, this plant also reduced the growth of native plants, this is a non-native plant which cause hinderance to the growth of native-plants besides damaging houses and hillsides. Similarity water hyacinths are floating plants with waxy green leaves were brought from Argentina to United States in 1880 for exhibition. People took plants home and intentionally put them into lakes. Results reveal that within a period of ten years, this prolific weed had spread across the South from Florida to Texas, completely covering the surface of many waterways, blocking boat traffic and smothering fish and native plants. So extinction is also caused not only by introducing non-native animals but due to non-native plants too. Kudzu-a-States and were brought from Asia and Argentina. These plants extended the growth of native plants.

(v) Predator and Pest Control - Some animal populations have been internationally exterminated because they are dangerous to humans and livestock. Animal control agents in USA trap or even shoot thousands coyotes bobcats, prairie digs because they are considered threat to people, livestock or crops. To lill 86,500 coyotes the cost involved is worked out by Office of Animals Damage Control amounted to $ 38 million. Animal rights activities oppose the killing of predators and comment that they have equal right to inhabit the landscape as humans. They suggest guard dogs predators thought to be dangerous. One may comment anyway he likes but one issue cannot be ignored that some animal population is exterminated.

(vi) Commercial Products - A variety of commercial products like furs, hides, horns and live specimens are being obtained from nature in addition to harvesting wild species for food. People consider it better and sustainable harvest compared to agriculture and industry. Despite international bans on trade in products from endangered species, man is involved in the smuggling of these products worth millions of dollars each year. Developing countries in Asia. Africa and Latin America with the richest biodiversity in the world are the main source of wild animals and animal products. North America and some wealthy Asian countries are the chief important.

Places below is information with regard to trade and commerce of animal and animal products :

a. The trade in wild species for pets is an enormous business. Worldwide some 5 million live birds are sold each year for pets.

b. Plants also are threatened by overharvesting. Wild ginseng has been nearly in many areas because of the Asian demand for roots that are used as an aphrodisiac and folk medicines.

c. In 1980, there were 1.3 million elephants in Africa. A decade later only 625, 000 remained. 1000,000 elephants were killed by poachers for ivory trade.

d. The population of African black rhinos dropped from 100,000 in the 1960s to about 3000 today due to demand of their horn. In Asia powered rhino fetches \$28,000 per kg because it has medicinal properties. In Yemen, a rhino horn dagger handle is sold up to \$1000.

e. Profits on wildlife smuggling is very high. Tiger of leopard fur coats can bring \$100,000 in Japan.

f. The united States imports 90% of all live cacti and 75% of all orchids each year.

g. European countries buy 3/4 of live birds.

h. Japan, Taiwan and Hong Kong by 3/4 of all cat and snake skin.

(vii) Hunting and Fishing - Overhunting and overharvesting is responsible for depletion of many species. Ectopistes migratorius in classic example which exterminated due to overhunting and habitat destruction. 200 years ago this was the world's most abundant bird with a population between 3 and 5 billion. It once accounted for about one quarter of all birds in North America. In 1830, John James saw perhaps a flock of billion Ectopistes migratorius. Between 1870 and 1890 hunting and habitat destruction caused the entire population to crash. The last known passenger pigeon was shot in 1900 and the last female bird died in 1914 in the Cincinnati Zoo. Overhunting cases also include extermination of great whales and American bisons. Bisons are killed for hides, tongues, food, clothing and shelter.

Studies reveal that 2.5 million great whales once inhabited the world's ocean. The blubber of whales was highly prized as a source of oil. Initially the slowest whales were caught by hand-thrown harpoons but in the 19th century with the introduction of steam ships and explosive harpoons it became possible to catch and kill even fastest whales. The International Waling Commission has prohibited catching of great whales. But countries like Iceland, Japan and Norway continue to catch whales under the guide of scientific research. They restore meat and blubber of these animals in their programme of scientific research and sell the same at a handsome profit.

It is a difficult task to count whales accurately. Should man catch whales is an ethical question. Do you think and find acceptable to hunt and kill large, rare, intelligent wild creatures like whales? Fish stocks have been seriously depleted by overharvesting. With the result, a huge increase in fishing fleet size and efficiency in recent years has led to a crash of many oceanic populations. Worldwide, 13 to 17 principal fishing zones are now reported to be commercially exhausted. Canada cut nets or seized foreign boats violating their territorial waters.

(viii) Habitat Destruction - Habitat is called living place of species. Many species are endangered by multiple threats to habitats by humans. Exotic species receive a greater threat to their habitants. For plants, trampling, grazing, collecting, flooding and other human actions constitute serious dangers. The current increase in extinction is habitat loss. Habitat fragmentation which means dividing and sub-dividing of dwelling places of species divides population of species into isolated groups that expose to catastrophic events. In a state of habitat fragmentation, very small populations may not have enough forests to breed adults under normal circumstances. This happened due to decrease in wooded area of Cadiz township in Southern Wisconism during European settlement. Saded area reduced miserably and drastically from 1831 to 1950. By destroying habitat, man eliminate not only prominent species but also many rare once of destruction of forests, wetland, and other biologically rich ecosystems around the world. Endangered species are kept in Zoological Parks. Zone are just jails additional solicitor general Altaf Ahmad told to Supreme Court days after 12 tigers died in Bhubaneswar's Nandan Kanan Zoo.

(ix) Tiger and Peacock Deaths - The death of 12 tigers at the Nandankanan Zoo musical Park in Orissa in July 2000 has raised serious questions about inbreeding over crowding professional management stocking of drugs, and availability of funds for the zoos in the country.

During the last six years, 44 tigers in Orissa, 18 Lions in Guwahati, 10 tigers in Hyderabad and six hunters in Patna - totalling 78 in all have died captivity. The death of the big cats are a lesson not only the zoo custodians but also for the scientific community and veterinarians. Doctors at the Orissa Veterinary College, Bhubaneswar, as well as senior forest officials in-charge of the zoological park, said the deaths could have been due to trypanosomiasis, a parasitic infection, which is endemic in nature and is spread by a particular type of fly.

All the victims belonged to group of 17 tigers which were administered the Beneril injection, a preventive medicine. Dr. A.T. Rao, Professor of Pathology in the OVC, who conducted the post-mortem on four tigers said that a possibility was that the preventive medicine was administered when the infection had reached its final stage. In July 2000, about 80 peacocks died in Morena district of Madhya Pradesh. Fifty peacocks were found dead at Maharajapur village and nearby areas and twenty-nine at Sitapur village in the district. The deaths were caused by the birds consuming bajra containing pesticides. The post-mortem report revealed that the death of 19 peacocks was due to the birds consuming foodgrains containing chemicals of he Organo-Phosphate category. The post-mortem indicated they had died due to hemorrhage of the stomach caused by chemicals belonging to the mercuric and carborate group.

Natural causes of extinction

99 percent of all species that ever existed are now extinct. Studies of fossils records provide this information. Most of those species were gone long before humans came on the scene. Species appear and disappear through process of mutation. Evolution can proceed slowly and gradually over millions of years. New forms of species replace completely lost. The do exist. Tiny Hypohippus has been replaced by a large modern horse but the genes of tiny Hypohippus perhaps still

survive on its distant offspring.

Mass extinctions have wiped out vast numbers of species even whole families. This happened when dinosaurs disappeared. An even greater disaster occured around 250 years ago when 2//3 of all marine species and nearly half of mass extinctions of species were :-

(i) Release of greenhouse gases in the atmosphere.

(ii) Large asteroids structuring to earth.

(iii) Catastrophies caused by climate changes.

Biodivesity in India

India's biological diversity is quite immense. She has over 45,000 plant Species which represented about 7% of world's flora and 6.5% of world's fauna. India is one of the 12 countries identified as mega-centres of biological diversity.

Embracing 19 biogeographic Zones and 25 biotic provinces, India incorporates an exceptional range of biodivesity, including all the world's major ecosystem. As one of the eldest and largest agricultural societies, India also has an impressive diversity of at least 166 Species of crop plants and 320 species of wild relatives of cultivated crops. Among the most neglected and threatened parts of India are its extensive coastal and marine areas. These areas are home to a high percentage of biodivesity, and the lifeline of at least 10 million fisherman and many million other residents

Biological Diversity Bill- 2000

The Biological Diversity Bill 2000, awaiting Parliamentary sanction and presidential approval, is a landmark legislation aimed at protecting India's bio assets and ensuring equitable benefit-sharing by the users and the owners.

Bill No. 93 for the year 2000 was tabled in the Parliament on May 15, 2000. Christened the Biological Diversity Bill 2000, it is intended provide for conservation of biological diversity, sustainable use of its components and equitable sharing of the benefits arising out of the use of biological resources and for matters connected therewith or in dental thereto. This preamble is same as the objectives spelt out the Convention on Biological Diversity Signed in 1992 at the UN Conference on Environment and Development at Rio de Janeiro. While 150 countries signed the treaty, the US was an exception, arguing that it will set back research efforts in biotechnology and protection of intellectual property. Subsequently, the US Congrees agreed to implement the terms of the treaty.

The various provisions of the Biological Diversity Bill, 2000, are as follows:

(i) Ecomarks - It is a labelling for environment-friendly household and other consume goods which satisfies the requirements of Indian standards for that product. The Bureau of Indian Standards is the implementing agency for this scheme.

(ii) Funds - Biodivesity Funds to be set up Central, state and local levels.

(iii) Authorities - National Biodivesity Authority to deal with request for access from foreigners; to grant approval for seeking IPRs in or outside India for any invention pertaining to be biological resource obtained from India. State Biodivesity Boards to deal with matters relating to access by Indians for commercial purposes and restrict activities violating objectives of conservation, sustainable use, and equitable sharing of benefits.

(iv) Safeguards and exemptions - Free access to Indians to use biological resources within the country for research. Use of bio-resources by vaids and hakims. Traditional knowledge to be protected.

(v) Objects and reasons - Regulate access to biological resources. Respect and protect knowledge of local communities. Protection and rehabilitation of threatened species. Involve self-government institutions in implementation.

Biodivesity Protection

Biological diversity is important and irreplaceable assets. Slowly and gradually we have become aware of the harm we have done and continue to do to wildlife and biological resources. Now we are adopting national legislation and international treaties to protect endangered species and biodivesity.

Hunting and fishing bill was introduced in the United States congress to protect American bison in 1874. This initiative remained ineffective because many legislators believed that all wildlife was so abudant that it will never deplete by human activity. But this feeling proved quite wrong. By the end of 19th century bisons had plunged from some 60 millions to only a few hundred animals.

By the 1980s most states had enacted hunting and fishing laws to restrict liberal shooting and poaching. The general idea being is to protect and conserve biodivesity and natural resources for future generation. The wildlife regulations and refuges established since that time have been remarkably successful for many species. At the turn of century, there were an estimated half a million white-tailed deer in the United States; now there are 14 million or more in some places that the environment can support. Wild turkeys and wood ducks were completely extincted some fifty years ago but now situation is different. They are now restored to several millions due to restoring habitat, planting food crops, building shelter or houses, protecting them during breeding season and other conservation measures.

The Endangered Species Act and CITES represent a new attitude towards wildlife in which we protect organisms just because they are rared and endangered. A few acts passed from time to time to preserve and protect species of diversity are stated below to save nature because saving nature means serving human population.

Endangered Species Act : U.S. Endangered Species Act of 1973 and COSEWIC 1976 provide powerful approaches to wildlife protection. The regulations under these acts identify endangered species and save biodivesity regardless of its usefulness to humans. The species under the Acts are classified into three sections.

a. Vulnerable Species - These species are naturally rare, or we can say that activities have put them to such a low level of numbers that their risk of being depleted is being felt. They are Bald eagles, gray volvas, brown bears and sea otters.

b. Threatened Species - These are those species that have declined significantly in total number and are on their verge of extinction. Spotted owl and gizzly bear are recognized in threatened species.

c. Endangered Species - These are those species which are considered in imminent danger of extinction. These worldwide endangered species are mammals, birds, reptiles and plants. The Untied States has some 1359 species in its endangered species $39 million per year is being spent to protect these species annually by USA.

In-situ Ex- situ Biodiversity Conservation

Many strategies have been used over the years to try to conserve biodiversity. They can broadly be divided into in-situ and ex-situ approaches. Both approaches are useful in biodiversity conservation. Ex- situ approaches involve collecting storing and conserving things away from the environment in which they were found. Thus seed collections, gardens, orchards and zoos all seek to preserve biodiversity in a safe and controlled environment away from where they were found naturally. While such an approach removes biodivesity from its threatened environment to safe area, it also removes it from the dynamic environment in which it was growing and evolving. Recreating the ideal conditions is an exaction task, and reproduction to propagate the species can be difficult or impossible. This form of conservation preserves species, but not necessarily habitats or ecosystems.

In-situ approaches concentrate on preserving and conserving bio-diversity within the environment in which it is found. In-situ conservation is the conservation of ecosystems and natural habitats and the maintenance and recovery of viable population of species in their natural surroundings and in the case of domesticated or cultivated species, in the surroundings where they have developed their distinctive properties.

In-situ conservation allows the ecosystem and species to continue to evolve within their environment, adapting and changing in response to changing environmental conditions. Thus the focus is on the habitat and ecosystem, as well as the species.

Although viable populations of some organisms can be maintained ex-situ either under cultivation or in captivity, these methods are far less effective than in-situ methods, and, generally, they are extremely costly. Likewise, although ex-situ methods are important under a number of conditions, in-situ methods are generally recognized as being more secure and financially efficient. The challenge in using in-situ methods is to expand our vision of protected areas to includes multiple use and extractive reserves and to develop new models for conservation including; Such innovative proposals as using damaged ecosystems to preserve rare, endangered, and threatened species; conserving Biodiversity in seed Banks and gene Banks ; Restoration on Bio-diversity : vegetation panting to control erosion : removal of contaminated soil : Fencing to prevent cattle : Reintroduction of exculpated species : imparting environmental education, Enforcing Environmental Legislation.

Advantages, Risks, and Opportunities

In-situ maintenance of biodivesity through the establishment of conservation and multiple-use areas offers distinct advantages over off-site methods in terms of coverage, viability of the resources, and the economic sustainability of the methods :

1- Economic sustainability - A country that maintains specific examples of biodivesity stores up future economic benefits. When the need develops and this diversity in thoroughly examined, commercially valuable genetic and biochemical material may be found. Establishing a conservation are in just not sufficient to assume its bio-diversity till protected automatically without risk. Many other risk both natural and man-made, remain. They are mentioned below :

a. Genetic uncertainty or random changes in genetic make-up due to genetic drift or inbreeding in-situ programs such as national parks, biological reserves, and other conservation areas appear to be somewhat favourable.

b. Natural catastrophes such as floods, fires, or droughts, which may occur at random intervals.

c. Demographic uncertainty resulting from random events in the survival and reproduction of individuals.

d. Environmental uncertainty due to random, or at least unpredictable, changes in whether, food, supply, and the populations of competitors, predators, parasites, etc.

2- Viability - Natural selection and community evolution continue and new communities, systems, and genetic material are produced.

3- Coverage - A worldwide system of protected and multiple-use areas would allow a significant number of indigenous species and system to be protected, thus taking care of the unknown until such time as methods are found for their investigation and utilization.

5

MARKET LINKAGES

The distance factor plays a significant role between the two main factors of the market, i.e., the producer and the consumer. Whatever goods the producer produces has to reach to the consumers who are located at distant places. For this he adopts various means and ways. These means and ways that the producer adopts to approach or establish links with the consumer are called as Market Linkages. The various means or modes of transportations like head load, roadways, railways, ropeways, pipelines, waterways, airways etc. serve as the main forms of market linkages.

Nature of Linkages

The market linkages are instrumental in the movement of goods from the producers to the consumers from the production of raw materials to their movements to the factories and markets or from the factories to the markets or from the markets to the consumers. In every stage, some of the other means of transportation is necessary.

Functions of transport in support of commerce

The development of the various modes of transportation or linkages affect expansion of trade and widening of the market. The entire economic and commercial life of a country depends upon services of transport. In fact, at every stage in the activity of production and marketing of goods some or the other mode of transport is needed. It is an element of cost, both in the production as well as in the distribution or marketing of goods, as the cost of transportation is included in the sale price of the product. Sale price of the goods affect both the producers and the consumers. Therefore knowledge of transportation is must for the selection of best and suitable transport services.

A planned approach towards development of transportation is essential for the industrial and commercial development of a country. As large markets, mass production and specialisation in industry and commerce can be made possible with its help. It also enables numerous goods, irrespective of size, weight, etc. to enter in the market to satisfy ever-increasing and varied consumer demand both on national and global level. Further, it also reduces the cost, increases the speed and safety in the physical distribution of goods.

1. Transport and Consumption - Efficient and cheap transport reduces cost of production as well as cost of marketing. Hence, it can reduce the prices charged to the consumers. Lower prices help quick distribution of goods and encourage mass consumption.

Supply of goods can be equalized from place to place and through proper adjustment of supply with the changing market demand, we can have stable prices. A single price can be established for a commodity all over the country. Transport creates place utility, e.g., sand at the seashore has no value but in the city it has a price. Complete system of physical distribution involves two complementary functions, viz, transport involving the creation of place utility and storage or warehousing involving the creation of time utility. Both these functions of transport and storage must be closely coordinated in modern marketing and both together shall determine the cost and efficiency of physical distribution for consumption or use.

2. Transport and Exchange - Transport is a major factor in marketing, it helps exchange, i.e., buying and selling of goods. It promotes competition in commerce. It widens the market and allows each area or region to produce only those goods in which it enjoys superior advantages, e.g., Bengal has specialised in the production of jute and jute textiles for the world market, U.P. in sugar, Assam in tea, M.P. in wheat and cotton, Maharashtra in cotton textiles, etc. Such areawise specialization of production is possible due to easy transport facilities. International exchange of goods depends upon the best means of transport. Eighty years ago, Mumbai could get coal more cheaply from South Africa than that from Raniganj.

3. Transport and Production - At every stage in the production of goods we require some kind of transport. Production for the market whether agricultural or industrial, depends upon the means of transport. Adequate and cheap transport facilities can reduce the cost of production of goods, and can enable production of goods for distant markets also. Transport helps considerably agricultural, mining and manufacturing operations.

Services of Transport

Services of transport are summarized below:

1. It provides employment to a very large population.

2. Its equalizes supplies place-wise and provides stable prices in all markets.

3. It creates place utility.

4. Transport widens the market.

5. Defence demands adequate and efficient transport.

6. It spreads culture and civilization promoting higher standard of life. We can have close social and cultural contacts.

7. It increases localisation of industries.

8. It helps agricultural and industrial marketing.

9. It makes persons and goods mobile freely moving.

10. It helps quick industrialization of a country.

In short, we can satisfy our innumerable and varied wants through monetary exchanges very easily due to modern means of transport. If there be a strike in our transport services, our entire daily life will be simply paralysed. Transport is an essential public utility service and it is now a specialized and independent branch of industry. It is no longer merely and auxiliary agency or aid to trade. It has assumed national importance. All-round -planned industrial and commercial development of every nation primarily depends upon efficient and cheap transport- land, water and air transport.

Influence of Geographical and Economic Factors

Geographical factors, e.g., topographical features and climate, restrict the development of different means of transport and limit their use and efficiency. Mountainous regions like the Himalayas and the Rockies; ice-bound regions like Greenland, North Canada; marshy forested regions of the Amazon and the Congo basins; sandy dry desert areas of North Africa and west Australia- all these have continued to be poor in the means of transportation. Apart from unsuitable topography, climatic conditions, excessive cold climate and frozen lands do not facilitate transport. Stormy and cyclonic weather conditions and flood-prone areas restrict the use of all means of transport.

The development of railways depend upon physical terrain and the climate of a region. The construction of railways in the plains is easy and economic. But it is very difficult to build them in forest, deserts, ice-bound and hilly regions. In the mountainous areas, the cost of providing railway transport is always high because valleys are to be bridged and hills are to be cut into tunnels.

As per the schedule is concerned, among all the means of transports, railway transport is least affected by weather conditions and therefore sticks to a regular schedule. But it is different for road transports which are profoundly affected. Also affected are air and water transports.

Economically advanced and prosperous nations have the greatest facilities of transportation, while economically backward countries have poor transport facilities. The economic factors affecting water transport are the volume of traffic, port facilities and charges, location of refuelling stations, perishability of products, cargo handling charges. The size of ship and canal charges.

Co-existence of Different Modes of Transport

There are several modes of transport employed in the world of commerce. At present, the businessman has a choice of many different forms of transport for his goods-by road, by railway, by inland waterway, or by sea or by air. The reasons for the existence of variety in the forms of transport is that each mode of transport has its own peculiar advantages and disadvantages. In some cases road transport is to be preferred to rail transport; in others railway transport is superior to road transport. In some circumstances, goods may be sent by coastal steamer rather than by railway; for other goods sea transport may be the only suitable form of transport; in certain circumstances, a trader may prefer even air transport. Thus, the businessman makes his choice. In judging competitive efficiency of different means of transport and their coexistence we have to consider carefully the following points:

1. Availability - In the mountainous regions, e.g., in Kashmir valley or Simla, businessmen and have only animal or human beings as a mode of transport and, in foreign trade in the absence of common land frontiers, businessmen may have to rely only on sea transport. For reaching the interior and village population, we may have at our disposal only bullock carts or trucks. In India, air transport is not available for all markets.

2. Regularity and Punctuality - Railway transport usually provides a higher degree of regularity and punctuality and it is more dependable in this respect. Air transport may not provide expected regularity and punctuality, particularly in bad weather. In winter transport may not be available as water may be frozen in cold regions.

3. Flexibility - Road vehicles are extremely flexible. At present in the U.S.A., computerised routes and charts are prepared and these charts give the most direct, or most economical route to be followed. Railway transport is not so flexible.

4. Safety - During transportation, goods are subjected to many uncertainties and risks of loss. If a certain mode of transport provides necessary safety and security to goods, there will be minimum chances of damages or losses during transit. The businessmen prefer truck transport. Goods enjoy a damages or losses are leisurely in truck transport. Goods enjoy a higher degree of safety and security. Sea transport is not reliable.

5. Cost - Every person wants cheaper transport. Air

transport is very costly and it cannot be used for heavy goods. Cheaper and economical transport can reduce production and marketing costs and the consumer can buy goods at reduced prices. This means cheaper transport can expand our markets, increase our sales and profits. Transport cost plays a important role in commerce.

6. Speed - Quick delivery is essential, e.g., in perishable goods. Roses are exported by India by air transport as they are perishable and they must reach the market overnight.

7. Suitability - It must suit to our needs, e.g., buoyancy of the water makes the effective weight zero. Hence, sea transport is very suitable for heavy and bulky goods.

Water Transport

Since the earliest times, water transport has been availed of to carry men and materials. The earliest boats were made of different materials. Simple boats are still in use in different parts of the world. The range and importance of water transport increased when wind power was harnessed with the use of sails. Overtime increase in the size and complexity of sailing has facilitated the growth of trade. Traders from Egypt and the Mediterranean coast have moved towards south-east Asia for trade by boat.

The invention of the steam engine has brought a revolution in water transport. Steam gave greater power so that ships could travel at greater speed and carry larger loads. Ships were built of iron and steel instead of wood; and they were safe and efficient. Ships now use nuclear power. Bigger and bigger ships are built to serve different purposes.

Water transport is available for oceans, coastlines, rivers, lakes, and canals. The bulk of international trade is carried on by means of ocean transport. Coastal and inter-coastal traffic contribute to a great extent to international trade. Inland water transport has promoted inland trade, stimulated the paces of industrialization and assisted in the distribution of commodities on a larger scale. It is now the cheapest mode of transport because it does not incur the capital and maintenance costs of transport by land-by roads and railways.

Water transport is divided into three groups:

(i) Coastal shipping.

(ii) Inland waterways.

(iii) Ocean or sea transport.

Ocean Trade Carriers

The trade carries or merchant ships that are employed to carry passengers and cargo are:

1. Bulk-cargo carriers - Carriers or ships carrying dry and liquid bulk cargos have been continuously increasing in size to earn better and higher revenues and returns. This is necessary to ensure their operational viability. Bulk carriers which carry dry bulk cargo are available in various sizes. The most common among them are 40000 DMT size of carrier.

2. Tankers - Tankers are special ships, designed to carry mineral oils and petroleum products. Petroleum has become an important source of energy after the Second World War. Ocean-going tankers, therefore, were brought into use for mass-scale movement of oil from the areas of production to the areas of demand. Bigger and bigger tankers, known as super tankers, are now built.

3. Cargo boats - It is a steamer that performs the

functions of a liner as well as a tramp. This ship carries a large quantity of bulky goods and a limited number of passengers. Most of the ships are special purpose ships and have cold storage facilities. Cargo liners have specialized their functions; therefore, they are designed and fabricated to suit a particular purpose. The carriage of passengers give them additional income in the form of fares. Cargo liners are generally equipped to load and unload the goods. Sometimes they use crane facilities available at ports.

4. Tramp - The tramps are small cargo vessels which carry bulky goods. They are generally chartered to carry full cargo. But often take the cargo wherever can, and do not follow any definite route.

5. Liners - Liners are steamers which follow a definite route on their schedules, irrespective of whether they operate at full capacity or not. They carry passengers, mail and a small quantity of cargo. Modern liners offer many good facilities to passengers.

The Suez Canal

The Suez canal is one of the largest shipping canals of the world. It has been constructed by cutting across the isthmus of Suez to provide water transport facilities between Mediterranean sea and Indian ocean. It is the oldest oceanic canal and carries a lion's share of international waterborne commerce between the west and east.

The canal was opened for navigation in 1869. It is 160 kms long with a minimum width of 60 metres and a depth of 13 metres. The canal is at sea level throughout its length. The canal connects two important lakes the Great Bitter lake and Timsah. The canal can handle 49 ships per day with a tonnage upto 50,000. It takes nearly 13 hours to complete the journey through the canal.

The canal is not only link between the two seas but also a vital artery of world commerce. It has resulted in enormous saving in time, distance and money. It has resulted in enormous saving in time, distance and money. It has opened up a new route to Africa, Asia, Australia and New Zealand.

Advantages

1. Many ports of call and ports are developed along the Suez Canal route as a result of the construction of the canal, e.g., Aden, Karachi etc.

2. It helped Britain to maintain political control over her colonies and seas by dispatching her army and navy through the canal.

3. Revenue amounting to millions of dollars annually from toll is collected by Egypt. Egypt's economy is based on this canal.

4. The canal served as lifeline to Britain as the trade between Britain and Common Wealth countries of the East was carried by this canal route.

5. The canal reduced the distance as well as saved time for ships moving between the eastern and the western countries. Thus distance between London and Mumbai via this route was reduced by 6760 km and time saved was by 23 days. Distance between London and Singapore was reduced by 3535 kms and time saved was of 17 days. This route even brought the eastern coast of North America closer to the Far East Countries of Asia. The distance from New York to Hong Kong was reduced by 2290 kms. Similarly distance between New York and Mumbai was reduced by 4409 kms.

6. One more shipping route made available to western countries to trade with the countries of Asia, Africa and Oceania.

Future Potentials of other important canals

In these days of oil crisis, construction of canals has gained more importance. In order to conserve exhaustible fuel resources and to make distribution of goods more economical, construction of shipping canals, wherever possible, has become a need of the day. Such canals reduce distance of hundreds of kilometres to few kilometers. Construction of new canals such as Isthmus of Kra in Thailand will shorten the distance between Rangoon of Burma and Bangkok of Thailand by nearly eleven hundred kilometers. Similarly, there is a greater possibility of construction of one more canal which will connect Gulf of Aquba with Mediterranean sea. This will pass through Israel, benefitting her in the time of political crisis.

Kiel Canal

Kiel Canal is an artificial waterway and is located in western part of United Germany. It connects Baltic Sea with North Sea, reducing the distance of nine hundred kilometers to ninety kilometers. It has mainly benefitted United Germany. Though constructed for strategic purpose it has been now internationalized, i.e., open for all the countries since 1962. The ships passing through this canal mostly carry bulky goods like iron-ore and coal as well as forest products.

The Pnama Canal

The Panama Canal connects the Atlantic and Pacific oceans. It has brought the East and West of North America close to each other. The canal has provided the shortest route between ports on the Atlantic and those on the Pacific. It is 80 km long and 12-26 metres deep. The water level of the canal is higher than the sea level. Hence, six lock gates have been provided to maintain the proper level.

Advantages

The Canal has eliminated the long and difficult trip around Cape Horn and brought the West coast of the USA nearer to Europe by more than 8000km. The canal has brought the Western part of countries of 9. America closer to the Eastern parts of the USA and the countries of Europe. Nearly 50% of the total traffic through the canal is American. It has opened up a new route for Europe for going to Australia and New Zealand. The distance between Liverpool and Great Britain and Willington of New Zealand is reduced by about 640 km.

Disadvantages

1. There is lack of major ports and islands on the route.

2. The canal route passes through thinly populated hilly areas.

3. The system of locks, for maintaining the water level, is very inconvenient.

4. It is a narrow shipping canal of a width ranging between 92 meters and 309 meters.

Shipping Canals

Shipping canals are the artificial construction of waterways through a narrow strip of land lying in between two water bodies. Construction of such shipping canals shorten the distances among number of areas of the world, saving time and fuel, both. In the world, there are three important shipping canals today.

1. KIEL CANAL.

2. PANAMA CANAL.

3. SUEZ CANAL.

Panama Canal Route

This ocean route came into being in 1914 when

Panama canal was opened for the traffic. This route is of vital commercial importance to U.S.A. This route gets divided into many branches, one going up to the western coast of South America and the other to the western coast of North America. While crossing Panama from the pacific to the Atlantic, the route gets divided into further three branches; one branch runs to the northern and eastern coasts of South America, the second branch runs to the eastern coast of North America; and the third joints the South Atlantic ocean route to Western Europe, via the Caribbean countries.

The chief ports on this ocean route are Boston, New York Orleans, Seattle, Los Angeles, San Francisco of U.S.A., Valparaiso of South America, London, Liverpool, Glasgow, Manchester, etc. The Panama Canal route serves Europe, the eastern and western coasts of Canada and the U.S.A., the Caribbean islands, Central American countries and countries on the coasts of South America, Australia and New Zealand. The export of important minerals, such as bauxite, iron ore, copper, lead, magnesium, tin, zinc, and nitrates are facilitated from South America to U.S.A. and Europe. The Panama Canal route has helped to improve the economic conditions of Central and South American countries.

Suez Canal Route

The Suez canal route came into existence after the construction of Suez Canal through Isthmas of Suez, connecting Mediterranean Sea with Red Sea. It is a great international trade route, connecting the East and the West. The Suez Canal route has been of immense commercial and strategic value to European countries and the U.S.A. in maintaining trade relations with East African countries, south and south-east Asian countries, the Far East and the Middle East. The canal route was a vital link between European colonists and their poor Asian and African colonies- a link between the modern Western civilization and the old Eastern civilizations. Trade in agricultural products and minerals to feed industries in Europe and the quick supply of finished products in the markets of the colonies continued through this route till the middle of the 20th century. New systems for defence, administration, education, cultural ideas, transport, business, banking, recreation, etc., were promoted and transferred by this route from the West of the East. It assisted the Europeans to become industrially, economically and politically strong.

Due to this canal route the distance between India and England was reduced by about 9,6000 kms. This route also eliminated the trip around to Cape of Good Hope. However, it was closed from 1967 to 1975 because of Arab-Israel war. Fuel supply on this route has never been a problem. The important coal refuelling stations on this route are Gibralter, Port Said, Aden, Colombo, and Hong Kong. As regards oil refuelling stations, Indonesia, Sarwak, the Middle East, the Black Sea, and the North African countries are the important countries on this route.

Most of the ports on this route are well-equipped with modern equipments for loading and unloading operations. These ports are: London, Southampton, Liverpool, Glasgow, Le Harve, Mumbai, Colombo, Hong Kong etc. The first division of traffic is related to the trade of Mediterranean countries with each other. Food and raw materials are important by France, Italy; oil from the Middle East by the West; cotton, tobacco, chromite from Turkey; and grains, oil, manganese from the C.I.S. The East-bound traffic is generally of manufactured goods.

The second division of traffic is relates to the trade between West European countries and Mediterranean countries. The commodities like fruits, wines, cotton, corks, silica, sulphur, etc. move towards Western Europe; while coal, forest products, machineries, tools, textiles and manufactured goods move towards the Mediterranean lands.

The third division of traffic relates to the trade among North America and East Africa and Asian countries from the eastern coast of Canada and the U.S.A. There has been a considerable flow of traffic through the Suez between these two ends. Rubber, tin, manganese, tea, tobacco, spices, coffee and jute products- are the major commodities which are carried towards North America Automobiles, machineries, petroleum-products and other manufactured goods move of East African and Asian countries.

The fourth division of traffic relates to the trade between West European countries and East African and south and south-east Asian countries. West-bound export consists of a variety of food products, agricultural products, minerals, oil and other industrial raw materials. Eastbound export consists of iron and steel products, locomotives, machineries, tools, chemicals and a variety of manufactured goods.

The fifth division of traffic relates to the trade between North America and Western Europe in the west and between North America and the Far east, Australia and New Zealand in the east. Japan and the Philippines have trade links with West European countries and North America. Rubber, tobacco, spices, tea, jute, sugar, wool, animal products, manganese, etc. move towards the European countries and the eastern coast of North America. Automobiles, petro-products, machineries and other manufactured products move from eastern North America and European countries to the Far East and Oceania. The trade of Australia and Now Zealand with European countries and North America is more on this route across the Pacific.

Pacific Ocean Route

The Pacific Ocean is the biggest water route in the world, and was developed mostly between western coasts of Canada and the U.S.A. and the Far East and Oceania. It was the last route to be developed because, of economic and physical conditions. The submerged islands, shoals in the ocean, cyclonic weather conditions and long vast stretches of water without land were the major problems. There water moreover, no refuelling centres for ship. Traffic on this route is poor.

The route links the eastern coasts of Asian countries-Japan and the Philippines- with the western coast of North America via Hawaii. A branch of this route form the western coast of North America goes south to Fiji, Australia and New Zealand. The far eastern countries though densely populated, do not have much purchasing power. Agricultural products are not grown in exportable surplus quantities. Japan in the east is leading in trade with U.S.A. and Canada on this route.

The important ports on the Asiatic side are Tokyo, Manila, Seoul, Yokohama, Kobe, Osaka. Shanghai, Hong kong and Canton. Those on the American side are Vancouver, Portland, Prince Rupert, San Francisco and Los Angeles. The U.S.A. and Canada export lumber, wood pulp, cereals, petroleum-products, minerals and various manufactured goods. Japan exports silk, tea, electronic goods, and automobiles. The other countries export sugar, rubber, tea, corpa, tobacco. Australia and New Zealand export minerals wool and other animal products.

South Atlantic Ocean Route

The South Atlantic ocean route does not lie exactly in the Southern part of the Atlantic Ocean. The route was established between European colonists and their colonies in Central and South America. European countries, Canada and U.S.A. were connected with Central and South American countries. The route did not develop in the southern part of the Atlantic Ocean because African and South American countries have little to exchange. Their geographical environments are similar; their products are the same; and both the continents are thickly populated.

The South Atlantic route has been developed by the Europeans and the Americans to import raw materials and seek markets for their finished goods. Therefore, the pattern of trade has been, till recently, more or less the same. Now the process of industrialization has begun in Mexico, Brazil, and Argentina. Agricultural products, pastoral products and minerals move from Central and South America countries including wheat, corn, coffee, sugar, cotton, tobacco, wool, vegetable oils, fats, waxes hides and skins, iron ore, bauxite, manganese, tin, oils, copper, etc. , to European countries, Textiles, iron and steel, medicine, paper, wines, electrical and electron goods- these are exported by Canada, U.S.A. and West European countries to Central and South American countries.

European, American and Canadian ports have contributed to the development of trade on this route. But there are only few ports in South American countries. Kingston, Havanal Pernambuco, Santos, Sao Paulo, Rio de Janeiro, Montevideo and Buenos Aires are the major ports through which trade is handled by the South and central American countries. The volume of trade on this route is comparatively small.

North Atlantic Ocean Route

The North Atlantic ocean route has been the busiest among all for more than a century. It lies in the northern part of the Atlantic Ocean and links Western Europe with the eastern coast Of Canada and U.S.A. Nearly 25 percent of the world's vessel tonnage is handled on this route. The countries on this route are the leading producers of the wide variety of goods. They exchanges good in large quantities. The volume of trade between them is always large. Passenger traffic on this route has also been regular because of old cultural ties. The world's fastest and largest ships operate on this route. People from either side enjoy a high per capita income and a fairly high standard of living.

The North Atlantic route is a great trunk route with a number of branch lines. The terminal branches extend from the Mediterranean to the North Sea in Europe and from New Found land in Canada to the Caribbean Islands and Central America. The North Atlantic route is the shortest route between Western Europe and North America.

There is no shortage of fuel on this route. Cheap and abundant supply of coal is available on both ends of the route. Oil-burning ships from North America carry enough oil for their return journey because of the high oil cost in Western Europe.

As large number of natural ports have deep waters, well protected sites and highly developed integrated hinterland. Most of the ports are free from ice during the major part of the year and they are well equipped with modern facilities for bulk cargo handling. The important ports on the American coast are Halifax, St. John, Montreal, Quebec, Boston, Baltimore, New York,

Charleston, Philadelphia and Noe Orleans; the important ports on the European coast are Liverpool, Glasgow, Cardiff, Southampton, London, Le Harve, Amsterdam, Bremen, Rotterdam, Hamburg, Antwerp, Oslo, Stockhol, Bordeaux and Lisbon.

Forty percent of the world's trade is handled on this route. The passenger and mail traffic has decreased recently because of the development of air transport. The shipments from Europe to North America mainly consist of manufactured goods, cars, wines, paper pulp, dairy products and luxury goods, while those from North America to Europe consist of iron ore, forest products, dairy products, grains, cotton, meat, petroleum and petro-products, machinery, vehicles, fruits, etc.

Geographical conditions are generally very favourable for navigation on the North Atlantic route. But the ice-bergs and dense for near New Foundland has prevented this route to follow great circle route. Cyclonic weather in winter also pose some problems for the ships on this route.

Cape Route

The Cape of Good Hope route is the world's oldest route which links West European countries, the Southern and South-East Asian countries, the Western and South African countries, as well as Australia and New Zealand. European colonists established their colones in Asia and Africa travelling by this route. They traded in industrial raw materials and agricultural products via the Cape route. They sought good markets for their finished goods; and spread modern ideal culture, etc. Political dominance over Asian and African countries was established by Europeans by travelling through this route.

Refuelling along this route is done at Madeira and

the Canary Islands. Colonists tried to develop new ports in their colonies. The important ports on this route are London, Liverpool, Cardiff, Swansea, Le Havre, Libson, Cape Town, Port Elizabeth. Durban, Dar-es-Salaan, Mombasa, Colombo, Mumbai, Adelaide, Melbourne, Sydney and Brisbane.

This route declined in importance when the Suez Canal was opened in 1869. But is promoted a vital trade link between the East and West till then. The importance of this route increased again following the closure of the Suez Canal in 1967.

Copper, gold, iron ore, uranium, cobalt, tin, lead, zinc, bauxite, diamonds, chromite, etc., and agricultural products, such as tea, sugar, cotton, jute, jute goods, hides and skins, coffee, wool, tobacco, cocoa, rubber, limber, vegetable oils, etc.- all these goods moved through this route towards the European countries and U.S.A. The east-bound movement was of manufactured goods, such as textiles, chemicals, medicines, drugs, electrical goods, machines and tools, transport equipment and a variety of manufactured goods. The traffic on this route declined following the opening of the Suez Canal.

Charter Party

It is an agreement whereby the shipowner agrees to let the entire ship or a part thereof to any person, called the Charterer, for the carriage of goods on the specified voyage of during a specified period of time for a sum of money which the Charterer undertakes to pay as freight for the carriage of the cargo.

The form of Charter Party contains the following information:

1. The lay-days. The Charter Party also contains

terms and conditions on which the ship-owner agrees to charter the ship.

2. The time required for the voyage.

3. The freight agreed upon.

4. The list of cargo specifying the goods by number, weight and marks.

5. Ports of loading and discharge.

6. Registered tonnage.

7. Name of the Charter of the shipper.

8. Name of the ship.

Contract of Affrerightment

A consignor who wants to ship the goods by sea has to enter into a contract with the ship-owner unless he owns the ship himself. When the ship-owner agrees to carry the goods by sea or furnish a ship for the purpose of carrying the goods in return for a sum of money, such a contract is called the Contract of Affreightment, and the sum of be paid is called the Freight. The cargo of a ship is also called Freight.

A contrast of Affreightment may be embodied in a document called the Bill of Lading, or in another document called the Charter Partly. The Bill of Lading is issued by the ship-owner to the shipper when he agrees to carry goods which forms a small part of the entire cargo. When the agreement is to carry a complete cargo or to furnish a substantial part of it, the Contract of Affreightment is embodied in a document called the Charter Party.

World Shipping Routes or Ocean Routes

The ocean routes, generally known as international trade routes, are influenced by geographical and economic conditions. These are:

1. The Pacific ocean route.
2. The South Atlantic ocean route.
3. The North Atlantic ocean route.
4. The Cape route.
5. Canal routes and canal charges.
6. Refuelling facilities.
7. Availability of ports and harbours.
8. Weather conditions.

The Suez Canal route and the Panama Canal route are also considered as international trade routes.

Containerization

The term containerization refers to the shipment of goods in uniform, sealed and reusable containers. This is a recent development in the handling of shipments in large quantities. It is damage free, economic and fast across land and sea. Recently, shipment in containers was introduced in air transport. The container facilitates the combined and coordinate-carrier cars are especially designed and fabricated to carry very large and bulky goods. Containers are made of waterproof and fireproof materials. The container are made of waterproof and fireproof materials. The container service is mainly for such commodities as perishable, chemicals, and drugs.

The consignor enjoys the benefits of container service, which are: reduction in the packaging cost; less likelihood of damage to goods; fast delivery; door-to-door service, and unit-load movement from factory to warehouse; no chance for rough handling of goods; loading and unloading by mechanized means; and safer and economic.

Container transport service is not useful for all kinds of goods. Coal, ores, steels, fertilisers, grains, jute, jute goods, hides and skins, etc. These are some of the commodities which cannot be transported by containers. Engineering goods, cotton textiles, manufactured goods, frozen foods and well-packed commodities are generally considered suitable for container transport.

Container transport service is widely used in industrially advanced countries. U.S.A., CIS, Japan, and the European countries often use this method of transportation, It would be worthwhile if all forms of transport adopt uniform, standard-size containers. Each mode of transport can then offer the best possible service, even if there is a break-of-bulk in transit. Container transport service is now becoming popular in developing countries. India introduced a railroad coordinated system in the sixties. Some marketing and sales organisations and Government undertakings have started using the container service very progressively. Oils, chemicals, liquefied gases and milk are transported in containers by road and rail.

The use of containers by Indian merchants, business houses and shipping companies is increasing day-by-day. At present, the facility of container cargo is provided at the ports of Mumbai, Chennai, Kolkata, Haldia and Cochin. The cargo-handling charges at Indian ports are generally high. Secondly, congestion at the lort in another problem. Mountains of cargo eat up the precious dock space in India. The Container Corporation of India is planning to introduce some measures to relieve the congestion such as:

1. The Inland Container Depots and Container Freight Stations are to be set up at major ports. The Wadi Bunder Container Terminal is the best example of the port side container terminal at Wadibunder in Mumbai.

2. Modern loading system to be employed to double the speed of operations.

3. Many of the port functions are to be taken out of the port areas.

Chief Trade Carriers

The trade carriers or merchant ships that are employed to carry passengers and cargo are:

1. Bulk Cargo Carriers - Carriers or ships carrying dry and liquid bulk cargos have been continuously increasing in size to earn better and higher revenues and returns. This is necessary to ensure their operational viability. Bulk carriers which carry dry bulk cargo are available in various sizes. Thc most common among them are 40000 DMT size of carrier.

2. Tankers - Tankers are special ships, designed to carry mineral oils and petroleum products. Petroleum has become an important source of energy after the Second World War. Ocean-going tankers, therefore, were brought into use for mass-scale movement of oil from the areas of production to the areas of demand. Bigger and bigger tankers, known as super tankers, are now built.

3. Cargo Liners - It is a steamer that performs the functions of a liner as well as a tramp. This ship carries a large quantity of bulky goods and a limited number of passengers. Most of the ships are special purpose ships and have cold storage facilities. Cargo liners have specialized their functions; therefore, they are designed and fabricated to suit a particular purpose. The carriage of passengers give them additional income in the form of fares. Cargo liners are generally equipped to load and unload the goods. Sometimes they use crane facilities available at ports.

4. Tramps- The tramps are small cargo vessels which carry bulky goods. They are generally chartered to carry full cargo. But often take the cargo whereever they can, and do not follow any definite route.

5. Liners - Liners are steamers which follow a definite route on their schedules, irrespective of whether they operate at full capacity or not. They carry passengers, mail and a small quantity of cargo. Modern liners offer many good facilities to passengers.

Communication

The word communication is derived from the Latin communis which means common. We attempt to communicate, that is, to establish a commonness with another person. There are three essential parts of communication. They are the source, the massage and the receiver. True communication takes place only when the massage means the same thing to both the parties.

The effective communication occurs when a sender sends a massage and a receiver responds to the message in a manner which satisfies the sender. Both must have identical meaning of the message.

In any exchange activity, communication is absolutely necessary. You may have the best product, package and so on. It may have fair price. But people will not buy your product, if they have never heard of it, or if they are simply unaware of its existence. The marketeer must communicate to his prospective buyers and provide them adequate information. People must know that the right product is available at the right place and at the right price. Here, comes, the need of proper communication system which can move forward a product, service or an idea in a channel of distribution. Through this, the marketer, informs and persuades buyers to accept, recommend or use the product, service or idea which is being promoted. Both the transportation and the communication are complementary to each other and serve the world of commerce as prime aids to the trade. They combat the world of distance by shortening them.

Modern Communication facilities

The present state of commerce can be mainly attributed to the role played by various communication modes. There have been rapid advancement in the development of communication systems and their services. The development has come as a boon to the modern commercial world.

The modern communication systems have been playing very significant roles in present human societies. Besides all modes of transport services that man has been using for last so many centuries, it is science and technology that have enabled man to transmit his messages in words and pictures to any corner of the world within a moment and at any time. Post and telegraph service has been the oldest among all. Telephones, radio, television, teleprinter, fax and telecommunication through satellite have become very popular during twentieth century.

The postal services have been in use from centuries ago. But an organized and government controlled postal service was introduced by Europeans. It was by British colonists in India. Telegraph service was introduced in second half of the last century in Europe, mostly attached to the postal services. Today it serves the largest number of people and covers the longest area. It depends upon other means of transportation, however it is the cheapest Post and telegraph services have played a great role in establishment of trade links in the world.

Telephone service cause in existence in fourth

quarter of last century. A direct conversation through wire could establish a very quick and assured link between two persons. This service became very useful during war times and other emergencies. Telephone service could not reach to rural areas except in developed countries. Teleprinter service in newspaper agencies while telex in industries and business are the improved facilities to get machine typed matters within few moments. Fax is a further improved facility to get photocopy of the printed massage at other destination.

Radio service and then television made a real revolution in communication systems in the world. Not only geographical but all other barriers are crossed by this new means of communication systems. The world has shrunk into a small place. People of different culture and taste have come closer and closer. These services have enabled man to established links with people of remote areas. They propagate and circulate news, educational programme entertainment and advertisement to establish trade links. Also the informations about any warning of naturally calamities like earthquake, volcanic eruptions or floods can be spread rapidly and thus the relief work. Radio service has been of an eminence use to military and in border areas.

The satellite communication of enormous significance since last few decades. At least 200 satellites have been launched in the space by various counties in the world for communications to receive different informations and relay the massages, news, programmes, sports, and business news, etc. Weather forecast, resource exploration and spying are some other jobs performed by satellite for other countries. Telecommunication service is very common in developed countries like the U.S.A., European countries,

Canada, Australia and Japan.

The communication services are poorly developed in developing and backward countries of the world because of-\

1. Less demand of such services because of poor development.

2. Lack of scientific and technological know-how.

3. Economic and cultural backwardness.

Air Transport

In recent years, air transport has been assuming increasing international important although it has not yet been able to render all types of commercial services. The two world wars gave a great impetus to the development of aerial transport.

The supreme advantage of air travel is its speed. Aeroplanes have been travelling with a speed greater than the sound and for this reason air transport is eminently suitable for passenger and postal services. During 1940-80, the number of passengers carried by airlines increased over forty times and the amount of mail by twenty times in many advanced countries like U.S.A. and the U.K. The advent of Jumbo Jets since 1970, has further revolutionized the air transport all over the world.

Market research on air freight costs conducted by Air Canada was published recently and has proved the fact that air transport is practicable and economical method of transport for many transport is advantageous only for passengers, mail, perishable goods and light goods having high value. Fruits, tea, drugs, vegetables and many other perishable goods have entered international market due to air transport. Even valuable components of machineries are being carried by air transport.

Advantages - It is most suitable for carrying goods of goods high value such as diamonds, bullion, perishable commodities which necessitate rapid transport, Customs formalities can be very quickly complied with and thus delay in obtaining clearance is reduced, It is quite suitable for carrying goods with normal bulk and in small quantities. It is most comfortable for passengers and safe for goods, It provides regular, efficient and quick service.

There are two types of air vessels- airships and aeroplanes. The airship is higher than air, while the aeroplane is heavier than air. Now we have jumbo jets. It can carry 362 passengers and can travel at about 2000 kilometres per hour. Supersonic Anglo-French Concorde is recently introduced which can travel at a speed of about 3,000 k.m. per hr. During the last fifteen years, commercial and civil air transport has gained considerable importance all over the world. Today, we are living in the space age and the Jumbo Jet Age. Businessmen, tourists and politicians have become reminded.

Disadvantage - It is costly for goods transport. Too many international restrictions and regulations over its services arrest the growth of air network in the world, Air transport is controlled to a great extent flying operations temporarily and cause uncertainty in its timetable, Chances of breakdown and accidents are frequent owing to the fact that all the technical aspects of air transport have yet to be exploited, It cannot carry bulky commodities on a large-scale and hence, it is not of much use to the mercantile community, Costs of operation and maintenance of aerodromes and ground organization are heavy. Its carrying capacity is not very large.

(i) India - Among the developing countries, India has acquired an important place in civil aviation. Her midpoint location on the routes between the East and the West gives her a key position in international aviation. The Indian Government, therefore, has been taking a keep interest in developing air transportation in the country. The air transport industry was nationalised in 1953 by the Indian government. And it has established two corporations: The Indian Airlines Corporation and Air India. The Indian Airlines Corporation is for traffic within the country and with the neighbouring countries. Air India operates flights at the international levels. Since the corporations were formed, there has been a rapid progress. Both airlines operate their scheduled services on various routes, linking major Indian cities and towns with those in the world. The process of modernization has been continuous and modern aircrafts are purchased replacing the old. Boeing 747 are operated by Air Indian and A-310 AIRBUS by both Air India and the Indian Airlines Corporation. There are about four more domestic airlines.

The Indian Airlines Corporation operates regular flights between important cities - Mumbai-Delhi, Mumbai Chennai, Mumbai-Kolkata etc. Air India has air transport agreements with a large number of countries, and is among the ten leading airlines in the world. It runs regular flights on the principal routes-Mumbai-Nairobi, Mumbai-Delhi-London-New York, Kolkata-Singapore-Tokyo, Kolkata-Jakarta, etc. A well-organised network of airports and air routes is very essential for the expansion of air traffic. About 85 airports are maintained by the Civil Aviation Department of the Government of India.

(ii) Asia - Most of the Asian countries are linked by

air from the east to the west from the north to the south. Almost every Asian country has tis own airline, providing international air services. Three-fourth of the Asian air traffic is by south-east Asian countries. Japan and India together, have more than three-fifth of the traffic in south-east Asia. China, another Asian country has comparatively less air traffic at the international level.

(iii) Africa - Many African countries are linked with North America vis Europe or South America. The route via Europe is preferred because there are more halts and, therefore, greater traffic. Many routes link African countries with Europe because of colonialism in the past. A route from North-West Europe links Tunis and Algier via Madrid and Rome. Other route links European cities with south cities on the Atlantic coast and the eastern coast of Africa. Cairo in Egypt is a nodal point for all international airlines. Most of the African countries have their domestic airlines. A few have international airlines too.

(iv) South America - International air traffic is (v) Europe - The second large volume of air traffic in the world is in Europe. Every major industrial city and capital is linked by airlines. Passenger traffic and mail service by air are quite regular. European airways account for about 25 percent of the world's air traffic. Most of the European countries operates one or two international airlines. The U.K. has three airlines, British Airways is the most important airline. Germany's Lufthansa, Switzerland's Swiss Air, France's Air France, the Netherlands's KLM, Belgium's Sabena, Italy's Alitalia are the world's leading airlines operating in the European countries. Scandinavian Airlines services is jointly operated by Norway, Sweden and Denmark. The

European Airlines have links with the entire world towards the east and west, North and South. International air-traffic in all European countries is of greater importance than the domestic traffic, because every European country is small in size and has an adequate network of roadways, railways and inland waterways.

Fifteen independent countries after the former Soviet Union split up all could not establish their airlines independently. The former Soviet airline, now the Russian Airline Aeroflot is operated all over the world. The other countries such as Ukraine, Urbekistan, Georgia etc. will soon come in the international scene. The capital leading commercial centre of Russia, Moscow, is linked with all the major international centres and capital cities.

(vi) North America - The greatest volume of air traffic is in North America. The busy mode of life, level of development and high standard of life may be the reasons. Regional and feeder, inter-continental and trans-continental air services are easily available in North America. There has been a dense network of air routes over the entire territory and the U.S.A. International routes operate from the Pacific coast to the Far East and Oceania and from the Atlantic coast to Europe, Asia and Africa. The busiest air traffic is between north-east U.S.A. and Western Europe across the North Atlantic.

The U.S.A. alone has 9,000 airports and more than 60,000 registered civil aircraft. The country is served by four large corporations:

1. The United Airlines.
2. Eastern Airlines.
3. The Pan-American Airlines.

4. Transworld Airlines.

Besides there are about 45 domestic air services. The chief commodities moved by air are light machinery, electrical and electronic goods, meals, liquor, fish, fruits, vegetables, fish etc. Canada has a dense network of air route in the southern part of the country. The domestic airlines operate east-west in the agricultural, industrial and mining areas. The leading airline is Trans-Canada Airlines.

CIS

The CIS made remarkable progress in making high power TU-104 jets and TU-114, the turbo-prop aircraft. Greater attention is concentrated on the development of big sized speedy aircraft with a huge carrying capacity. A large number of air routes link Moscow with major international centres and capital cities - London, Warsaw, Buddhapast, Berlin, Parague, Sofia, Bucharest, Beijing, Havana, New Delhi, etc.

6

ENVIRONMENTAL POLLUTION

Introduction

The Planet Earth along with its living organisms and atmosphere which sustains life is known as the Biosphere. The different components of the biosphere interact in a delicately controlled and orderly manner and it is this factor that ensures the survival and growth of the biotic elements or living beings. Any adverse effect of disturbance of the delicate balance of the biosphere caused by man's careless or excessive exploitation of natural resources, can lead to the by man's careless or excessive exploitation of natural resources, can lead to the creation of unfavourable conditions for the well-being or even survival of mankind.

1. Environmental Degradation - Environmental degradation means deterioration of the environmental quality at global, regional and local levels by both-natural processes and human activities.

2. Environmental Pollution - Environmental pollution can be explained as lowering of the quality of environment of local level caused by human activities for exploitation of resources.

Pollution

Pollution indicates the adverse changes brought by human activities in the environmental quality at local level. Sometimes, the effects of human activities are so wide that the environment is degraded not only at local level but at global and regional level too.

As such we define Pollution as an undesirable change brought in the physical, chemical and biological characteristics of air, water and soil that may harmfully affect the or create a potential health to any living organism. In this way, Pollution is direct or indirect change that has come in any component of the biosphere that is harmful to living organisms and man that has come. It has an adverse effect on cultural and natural resources or general environment.

Meaning of Pollutants

Pollutants can be defined as any substance which causes lowering the quality of environment or pollution. A pollutant is any solid, liquid or gaseous substance found in such a concentration that it may have adverse or harmful effect or may be injurious to the environment, Pollutants are the residues of things, which we make or use or throw away. We know that the rivers are polluted by wastes emanated from factories, air is polluted by the gas, smoke emanated from chimneys of the factories, industries and thermal power plants etc.

According to Smith Every human society, be it rural, urban, industrial and most technologically advance society dispose of certain kinds of by-products and waste. Products which are injected into the biosphere in great quantities and they affect the normal functioning of ecosystems and have an adverse effect on plants, animals and man are called pollutants.

Thus a pollutant can be as under:-

1. Indirect being mediated via his resource organisms or climatic change. We may illustrate it with the help of an example. A water pollution can mean anything put into water which was not there in its natural state. In this way, any substance can be a pollutant if too much of it is present having an adverse

effect.

2. Direct.

3. Any physical property e.g., noise, heat. It is released intentionally or inadvertently by man into the environment with actual or potential adverse, harmful or unpleasant and inconvenient effects.

4. Any Geochemical e.g., dust, sediment, substance, biological organism or product.

5. Any chemical e.g. organic or inorganic compounds, radionuclide, organophosphorus compound, or trace gas.

Classification of Pollutions

Pollution have been classified on various bases. Chief bases are as under:

1. Classification on the Basis of Cast Pollution - It is discovered by the research studies that billions of tons of municipal waste are generated each year. About 60-70% of it including paper, yard and food wastes are easily biodegradable. As such, it is especially amenable to biological treatment. Again, about 12-16 percent rubber, leather and plastic waste are potentially biodegradable. At present, merely a small faction of the potential waste is treated by utilizing biotechnology.

2. Classification on the Basis of Nature of Pollutants - Pollution is classified on the basis of the nature of pollutants. Pollutants are broadly classified into two categories as under- Biodegradable Pollutants, these pollutants are the domestic wastes that can be rapidly decomposed under natural conditions and create problems when they accumulate. Non Degradable Pollutants, these pollutants include the materials and poisonous substances which either do not degrade or degrade very slowly in nature such as aluminium,

mercuric salts, long-chain phenolics, etc. evidently they are not naturally cycled in ecosystem.

3. Classification of the Basis of Human Activities - Pollution is generally caused by human activities. On this basis pollution can be divided into two broad categories- Social Pollution is caused in different aspects of the society. It is on account of cumulative effects of extreme events/hazards and pollution. Physical Pollution is caused by human activities by human activities due to lowering of the quality of physical components of the environments.

4. Classification of Pollutions on the Basis of Environmental Types.

Types of Pollutants

A proper balance between these two factors (Resources production, Waste disposal) is necessary for a healthy biosphere. The resources are created with a useful purpose of some economic advantage for the human being. However on actual use it is observed that some such resources like synthetic detergents, persistent pesticides, medicinal drugs, non-biodegradable substance, etc., have highly serious side-effects. Some of these wastes are of such a nature as cannot be degraded or processed by nature within a reasonable period of time Consequently, such wastes begin to accumulate and pollute the environment. Some device is indispensable to convert such wastes into usable form. Such a device would not only relieve of pollution but also improve, or at least, maintain environmental quality and the production of future resources. Reversal of environmental deterioration is one of the major goals of our society.

Various types of pollutants are as under:-

1. Agrochemicals such as fungicides and fertilizers.

2. Photo chemical oxidants.

3. Radioactive waste.

4. Solid waste.

5. Gases such as nitrogen, carbon monoxide, sulphur oxides.

6. Deposited matter such as dust, smoke, tar, grit, etc.

7. Complex organic substances such as ether, acetic acid, benzene, etc.

Natural Degradation

Micro-organisms cause much of the natural degradation that slowly remediates contaminants released into the environment. The specific mechanisms involved include:-

1. The bioregeneration of granular activated carbon.
2. Formation of biological barriers.
3. Bioenhanced filtration.
4. Accumulation.
5. Biodegradation.

Genetic modification of the micro-organisms involved can sometime further improve the above methods. At present, attempts are being made to improve micro-organisms through selective breeding.

Water Pollution: Meaning and Causes

Water, is one of the most important and most precious of natural resources. A regular and plentiful supply of clean water is essential for the survival and health of most living organism. On account of rapidly expanding industrialization and excessive population growths, most of our rivers, lakes, streams and other water bodies are being increasingly polluted.

Water is regarded as Polluted when it is changed in its quality or composition, directly or indirectly as a result of mankind's activities so that it becomes less suitable for drinking, domestic, agricultural, recreational, fisheries or other purposes for which it would otherwise be quite suitable in its natural state.

According to P. Viver, "Water pollution is defined as a natural or induced change in the quality of water which renders it unsuitable or dangerous as regards food, human and animal health, agriculture, fishing or leisure pursuits."

According to World Health Organization, "Foreign materials either from natural and other sources are contaminated with water supplies and may be harmful to life, because of their toxicity, reduction of normal oxygen level of water, aesthetically unsuitable effects and spread of epidemics."

Causes of Water Pollution

Main Causes of water pollution may be classified under two heads and under:-

1. Heat - It comes from power and manufacturing plants using water for cooling purposes.

2. Radioactive substances- They result from mining and processing of radioactive ores and the use of radionuclides in industry, medicine, research and nuclear weapon testing etc.

3. Sediments or silt - i.e., particles of soil, sand and minerals.

4. Persistent substances - e.g., non-biodegradable organic and inorganic substances.

5. Groundwater Pollution - Groundwater is the water that naturally flows through and is stored in soil and rock bodies beneath the land. Groundwater

contamination occurs when liquids, usually rainwater move through waste disposal sites and into the groundwater, carrying pollutants with them. The resulting mixture of liquid and pollutant is called leachate. The groundwaters are threatened with pollution from seepage pits, refuse dumps, septic tanks, barnyard manures, transport accidents, etc., and with diverse agricultural, chemical of biological pollutants. One major that arises from the effluents that are discharged on the ground so as to prevent pollution in rivers and streams. Other important sources of pollution of groundwater are inclusive of sewage and soluble salts.

6. Mercury Pollution - Rapidly increasing amounts of metals have been discharged into the environmentally industrial establishments. Out of these, mercury has been one of the worst offenders as a dangerous pollutant. It is responsible for several deaths in Sudan and Japan. It has caused the Minamata disease in Japan. Various Swedish rivers and lakes have already become heavily polluted by the widespread use of mercury compounds as fungicides and algicides in paper making and pulp lumber industries. The problem of mercury pollution in Sweden has been studied by Ackfors with particular reference to conditions in the aquatic habitat. Fishes have been found to absorb and concentrate mercury, in the form of methylmercury, to dangerous levels.

7. Thermal Pollution- In the category come mainly Heat and radioactive substance. These are the wastes chiefly from thermal and nuclear power plants- which use large quantities of water. There are some other industries also that give out waste water after use. Nuclear power plants emit ratio nuclides. In our country the quantity of waste water in highest in the thermal power plants. After use at very high temperatures this

waste water in highest in the thermal power plants. After use at very high temperatures this waste water is drained out to the streams, rivers and lakes. It adversely affects the aquatic life in these water bodies. It is given the name thermal pollution because heat acts as pollutant. The nuclear power plants cause fall out problems as well as release waste heat. It contributes to thermal pollution. A great number of plants and animals are killed outright by the very hot water. The waste water from nuclear power plants has following adverse effects on quintic life:-

(i) It is studied that thermophilic algae, cyanobacteria and fungi have been isolated from many hot habitats in the past few years.

(ii) Many workers have enumerated the effects of thermal pollution on microbes, Plankton, plants and animals. It is shown by Vannote and Sweeney that adult body size and fecundity of several aquatic insects depend mainly on thermal conditions during larval growth. It was found when growing in non optimal temperature range, the larvae developed into smaller-sized and less found adults.

(iii) Effluents from thermal power plants sometimes exerts adverse affect on the fish population by inhibiting the spawning run which blocks the access of fish to spawning grounds. It is found in an study of the effects of thermal effluents on marine organisms in the USA, Warinner and Brehmer that carbon assimilation by natural phytoplanktons was affected by an artificially induced increase in water temperature.

(iv) Discharge of waste heat in rivers affect the indigenous fish population as well as their migration or spewing.

(v) It is studied that the rapidly spreading weed Elodea

condenses has become the dominant species at thermal outfalls and heated areas in Alberta and that in such areas this weed has tended to replace species of Myriopyllum.

(vi) Discharge of heated effluents affects species composition and production of submerged macrophytes, in a water body.

(vii) Heated water produces more serious effects in those cases where an important biological resources is diminished. Although fish and other swimming animals can move freely in an out of a thermal plume yet sessile, benthic invertebrates cannot do so. Therefore in their case a plume can exert a much more profound effect.

(viii) Warm water holds less oxygen and hence its BOD rises. Water ooC. has approximately 14 ppm of dissolved oxygen whereas that at 200C has only 9 ppm.

(ix) Significant shift is caused in algal forms and other organisms towards more heat-to learnt forms. It results in decrease in species diversity.

(x) It leads to early hatching of fish eggs.

8. Industrial Effluents - In this category come the effluents from breweries, tanneries, dying textiles, steel industries, paper and pulp mills, mining operations etc. These effluents are both, inorganic and organic pollutants. Among them are included oils, greases, plastics, metallic wastes, plasticizers, suspended solids, phenols, toxins, cynides, DDT; acids, dyes, etc. Many of those create serious pollution problems. Sulphuric acid waste from coal mines increase the hardness of water. It has naturally disastrous effects on living organism. The heavy metal effluents, discharged from industries, include Nitrates, Cadmium, Mercury, Copper, Chromium and lead etc.

9. Domestic Wastes and sewage - Discharge of untreated or partially treated sewage into rivers is main cause of pollution, especially near big cities. Waste disposal in many countries is still archaic. Discharge of treated or untreated sewage into water bodies can produce some or all of the following symptoms:

1. Stimulation of algae growth and also a shift in the algal flora to the blue-green algae, leading to production of obnoxious blooms, floating scums or blankets of algae, etc. Decomposition of such algal masses in turn leads to oxygen depletion. Sewage discharge into waterways can lead to the spread of waterborne diseases.

2. Depletion of oxygen content due to biological oxidation of organic matter.

Sewage is the waterborne waste derived from home and animal or food processing plants. Domestic waste includes human excreta, paper, cloth, soap, detergents etc. These are a major part of the pollutants that enter water. Uncontrolled dumping of wastes of rural areas, towns and cities into ponds, lakes, stream of rivers, is being observed. Due to accumulation of sewage and other wastes in these water sources, it does not remain possible to recycle it. As a result, their self regulatory capability is lost. Due to accumulation of sewage and other wastes in these regulator capability is lost. Due to higher levels of pollution the decomposition of these waste by aerobic microbes decreases. The water loses its self-purifying ability. Water is rendered unfit for drinking and other domestic uses. The decomposition of sewage and other wastes being largely an aerobic process, their accumulation in water increases its oxygen requirements.

Phosphates, the major ingredient of most detergents, leads to luxuriant growth of algae. It forms water blooms.

This extensive algae growth consumers most of the available oxygen from water. It becomes detrimental to growth of other organisms. It produces a foul smell upon decay. Some decomposing plants also grow. These plants are known to produce toxins as strychnine. The latter kills animals.

Effects of Water Pollution

1. Metals - Metalliferous mining, smelting, and refining release metals to the soil atmosphere, water and sediments, e.g. electricity generating plants, incinerators, sewage sludge, and residues from leaded gasoline. Studies are made to know the long-term effects of metal accumulation in the soil resulting from applications of sewage sludge.

2. Fertilizers and Detergents - Some of fertilizes and detergents are fairly persistent pollutants. A major ingredient of most detergents is phosphates. When discharged into water, these phosphates support luxuriant growths or blooms of algae. After dying the algae release their component inorganic and organic substances back into the water. Thus, they become pollutants themselves. Excessive growths of algae rob the water of its oxygen.

3. Silt - The dirt and dust particles which are carried from land to water. It constitutes a very serous pollutant. These particulates originate from the weathering and erosion of rocks. Silt particles are mainly composed of silicates and carbonates with some phosphates and heavy metals absorbed on them. Silting can choke irrigation channels and rives. Edible fishes, e.g., salmon, are unable to spawn on silted gravel beds.

4. Weedicides - Residues of various weedicides and insecticides often accumulate in agricultural; soils rapidly. Insecticides are designed to kill insects. Some

herbicides differ from insecticides in killing both desirable species as well as the intended targeted. Their impact may not be confined to the plant community or agricultural crop but they many also act indirectly through effects on such soil microbes as nitrogen fixing blue-green algae and bacteria. This in turn may cause harm to the growth and production of higher plants. The impact of such chemical wastes as phenols, acids, metals, etc. applied to soil can also be fairly high and may tend to complete destruction of vegetation and also soil sterilization.

5. Man's Activities - Almost all of man's activities produce wastes. Many wastes can be recycled or reduced at source. Yet there are but some wastes and their residues which are inevitable and have to be placed in landfall sites. Up to 80-90 percent of our refuse has generally been disposed off on the land. Hazardous wastes come from industries, agriculture, and other sources. In fact, much of a nation's prosperity and economic well-being originate from activities that have produced hazardous wastes. Their proper disposal often presents tricky problems and needs careful consideration. However, some big chemical companies are now installing huge incinerators with a view to handle the wastes produced by them.

6. Acid Mine Drainage - Acid main drainage is chief source of pollutant. The reason is acid generation and discharges continue even after mining has ceased. Coal mining operation releases some acid arising from the exposure of iron sulphide minerals found in the coal for the purpose of keeping control, certain steps may be followed during the mining operation to prevent the acid from being formed.

7. Mining Activities - Mining activities constitute one

of the largest producers of solid waste. Extraction of fuel being a destructive process, environmental degradation is an inevitable result of such mining. The various environmental pollution problems arise from mining. The chief is of solid wastes. The handling and disposal of the solid waste is the principal cause for disturbed lands. During mining and benefaction, environmental problems can arise from the runoff from the disturbed area, and dust form blasting, extraction, haulage, storage, and grinding, etc.

8. Municipal Solid Waste - It is reported that most manipulates face the problem of managing the ever-increasing volume of municipal solid waste in an environmentally acceptable manner. An integrated strategy using the Four Rs is regarded as the most efficient and cost effective way of reducing the MSW requiring landfill, yet it is not being done. So there is continued reliance on landfilling. Given the current landfill capacity crisis, incineration may gain support as an integral part of an effective waste management strategy.

Measure to Check Water Pollution

Some measure to check water pollution are as under:-

1. Detention and National Treatment - In recent year storm-generated pollution has become one of the greatest offenders in the area of water pollution. It can contribute heavily to BOD, COD total suspended solids, total N and orthophosphate loads of water. An effective means of managing storm-generated pollution involves source control of storm water runoff by detection and natural treatment in a lake. By using this procedure it is possible to bring about an immediate improvement of runoff quality and make it possible to remove pollutants

on a long term basis.

2. Reverse Osmosis - For a number of industrial effluents, reverse osmosis has proved to be very effective for reclaiming or concentrating specific chemical or materials. However, less attention has been paid to the use of reverse osmopsis as a municipal wastewater treatment technique where the concentrations of toxicants and heavy metals are relatively low. Reverse osmesis is found capable of removing essentially all viral and bacterial particles of cells as well as most suspended and dissolved inorganic and organic chemicals from water. This technique can be used for separation of heavy metals from water and secondary municipal effluents. The pesticides and organic phosphates can, some time, be completely removed from water by using this method. The removal efficiencies of various toxicants vary with their chemical nature.

3. Ozonization - It is found that ozonization is useful to disinfect drinking water in Switzerland and some other countries. Treatment of such polluted waters with ozone and do as under-

a. Improve its taste and odour.

b. Disinfect the water.

c. Oxidize of pollutants.

Thus, water is made suitable for drinking. Some of the pollutants are directly oxidized by ozone while others react slowly with ozone decomposition.

4. Various Techniques - Various techniques are currently being devised to control water pollution. The techniques aim at removing various chemical, biological and radiobiological pollutants through various methods as under:-\ a. Ion exchange and reverse osmosis.

b. Electrodialysis.

c. Physio-chemical methods as absorption.

In most such methods, the high costs constitute the main obstructions and are the limiting factors for general and wider application to control water pollution. The reverse osmosis can be used to separate any substance in liquid or gaseous solution. It involves the use of a porous membrane the chemical nature can be made such that it has a preferential attraction for the solvent and a similar repulsion for the solute. This technique was, in the beginning, developed to purify seawater but is now playing a significant role in the Waste reclamation, Water purification, Water rnovation and Water pollution control. The technique is found valuable in the treatment of hard waters rich in calcium and magnesium. Reverse osmosis can be used as an economical and effective method for upgrading sewage other countries. Treatment such polluted waters.

5. Efforts by WHO - The World Health Organization, has proposed International Standards for Drinking Waters on the basis of which five classes of quality as distinguished, viz., as under :-

1. Characteristics affecting the acceptability of water. WHO has prescribed the maximum permissible concentrations of various kinds of pollutants. The characteristics for water acceptability include such properties as colour, smell, taste, tepidity, pH, hardness, etc.

2. Specific chemicals affecting health.

3. Toxic substances.

4. Radioactive pollutants.

5. Biological pollutants, e.g., microbes and pathogens.

6. Suggestions Given By EFWP - At its 16th

Symposium held in Zurich in 1972, the European Federation for Water Protection has given its valuable suggestions as under:-

a. First, a greater importance should be given to water bodies, both flowing and stationary, used for recreation, nautical sports, bathing, fishing etc. This should be more particularly for their value as sources of industrial and drinking water supplies.

b. Secondly, the groundwater should be suitably conserved and protected.

c. Thirdly, it should be kept in mind that it is not only biodegradable pollutants that cause the major pollution load. Instead there are several kinds of ingredients such as heavy metals, refractory chemical compounds, mineral oil, exist which cannot be decomposed by mechanical or biological means; inorganic fertilizers and various biocides.

d. Fourthly, preventive measures against possible future detrimental effects of pollution are even mor important than the repair of damage already inflicted. Hence, further efforts should be made to reduce the amounts of solids and liquid wastes by reutilizing or recycling some of their components.

e. Fifthly, efforts for control of water bodies and of plants serving the purpose of water protection should be reinforced and carried out by all available means, including legal enforcement.

5. To Stabilize the Eco-system - The most reliable way to control pollution and improver water quality is undoubtedly to stabilize the ecosystem by balancing the input and output of energy and nutrients. Some reasonable ways to increase stability of aquatic eco-system are as under:-

a. Fish management and aeration.

b. Trapping of nutrients.

c. Harvesting and removal of biomass.

d. Reduction in waste inputs.

Air Pollution

It need nor iterate that no life is possible without air. A living being can survive for a few days without water and for a few weeks without food but can nor survive even for a few moments without breathing i.e., inhaling air. Our atmosphere is gaseous envelope which surrounds the earth from all sides. The air is composition of several gases mainly nitrogen, oxygen, argon and carbon dioxide.

Air pollution takes place through enrichment of our atmosphere or air with noxious gases and other undesirable substances, caused largely on account of burning of fuels and through the release of gases by various industries and automobiles, the burning of fossil fuels releases huge amounts of CO_2 and other gases.

Thus Air pollution can be defined as the contamination of air with dust, smoke and harmful gases. The air pollution is generally made though the pollutants of gases and solid and liquid particles of organic and inorganic chemical. The various kinds of gases such as carbon monoxide, sulphur dioxide, hydrogen sulphide emissions from volcanoes, swamps, dusts, salt spray, pollens from plants, etc. are continuously added to the air by natural processes. Air pollution can be defined as the state when the natural composition of air is distributed either by natural of man-made sources of activities or by both.

Position of Air Pollution

It is revealed that in industrially SO2 advanced

countries, sulphur oxides constitute a very harmful air pollutant. Excessive enrichment of air with these oxides can have grave, and even lethal, effects on human beings as well as animals and plants. $SO2$ is the most common pollutant in human beings as well as animals and plants. About 60 percent of the $SO2$ emissions come from power stations wherein coal is used. In North America, photochemical oxidants, sulphur oxides and nitrogen oxides are common air pollutants. It is also revealed that, now the major cause of air pollution has shifted from SOx to photochemical air pollutants. Researchers have revealed cases of serious vegetation damage, caused by smog, ozone, peroxyacetyl nitrate, fluorides or ethylene, and polluted air has been established as being responsible for adverse effects on forest yields.

Increasing quantities of air pollutants are now causing changes in the atmosphere. Through our emissions of CFCs and other compounds, making has disrupted the ozone balance in the stratosphere. Hence, several countries have signed an agreement-the Montreal Protocol- to reduce and eventually eliminate emissions of substances that destroy the ozone layer. The latest research findings indicate that the depletion of the earth's ozone layer is proceeding at a faster rate than expected.

Causes or Sources of air pollution

The causes of air pollution are as under:-

1. Man-made causes or sources - These causes include human activities, such as industries, factories, urban centres, aircraft, nuclear experiments, automobiles, agriculture, power plants etc.

2. Natural Sources or sources - These sources include volcanic eruptions, deflation of sand and dust, forest, or wild fires of natural vegetation, etc.

Main Sources of Air Pollutions

Main sources of air pollution are as under:

1. Textile Mills - Textile mills emit great amounts of cotton dust. Pandit et al., have highlighted the concentration of airborne cotton dust particles from coarse and fine cloth mills located in Maharashtra, Vidarbha and Gujarat, and recorded average dust contents of 176,274 and 763 mg/100 m3 for fine mills, coarse mills and ginning presses respectively. Cotton dust implicate in the disabling occupational disease known as Byssinosis which is widespread in textile workers in India.

2. Furnaces - Future projections indicate that coal will be utilized to produce synthetic fuels by gasification and liquefaction. These processes, in turn, may pose new toxicological problems. The aqueous condensates derived from coal gasification and diliquefaction processes constitute primary effluents. The major organic components in the condensates are phenols, cresols and some low molecular weight carboxylic acids.

3. Thermal Pollution - In our country, there are a number or thermal power stations and super thermal Power stations. The National Thermal Power Corporation is setting up four mammoth coal-powered power stations to add to the energy generation situated as under-

a. Farakka in West Bengal.

b. Ramagundam in Andhra Pradesh.

c. Korba in M.P.

D. Singrauli in U.P.

The coal consumption of thermal plants is several million tonnes. The main pollutants are fly ash, sulphur oxide and other gases and hydrocarbons. The are tree thermal power stations at the Indraprashta Estate,

Rajghat and Badarpur in Delhi. They add to the air pollution. The Indraprastha plant consumers about four thousands tonnes of coal and in the fire units. Badarpur, the largest, consumes daily about 10 thousands tonnes of coal.

4. Industrial Air Pollution - In our country, there is rapid development in the agricultural technology and engineering. There is also rapid industrial development.

Nitrogen oxides probably pose the most serious air pollution control challenge. These oxides come both from automobiles and from certain stationary combustion sources. About 95 percent of the NOx act photochemically with hydrocarbons and SO2 to form undesirable compounds, with a shift of residual No to NO2.

5. Automobiles of Air Pollution - At present there is a rapid increase in number of Automobiles e.g. cars, trucks, buses and two wheelers. It is has raised the traffic density in towns and cities. They pour tonnes of gaseous pollutants into the air daily.

It is reflected from the studies that ever increasing vehicular traffic density poses continuous threat to the ambient air quality. About a decade and a half there were over 300 million cars, trucks and buses the world over and their number is increasing rapidly., India 2.8 million vehicles of which more than 65% were two wheelers operating in petrol. There has been an addition of not less than thirty percent vehicles now.

In the major cities of the country about 1100 to 1500 tonnes of pollutants are being emitted into the air daily, of which 50% come from automobile exhausts. It is revealed by the cities that in the major metropolitan cities, vehicular exhaust accounts for 70% of all Co2, 50% of all hydrocarbons, 30-40% of all oxides and 30%

of all SPM. In Delhi, there are more than three million vehicles which are registered there as well as registered in other status but being plied there. During peak traffic hour automobiles of all classes emit as much as 900 kg of CO, 350 kg of hydrocarbons and 90 kg of nitrogen oxides.

Causes of Air Pollution

Main causes of air pollution are as under:-

1. Metals - These include nickel, vanadium, titanium, beryllium, arsenic, tin, lead, cadmium, etc. They have been detected in the atmosphere in the form of solid particles or liquid droplets or gases. Mostly they come from fumes and dust from metallurgical processes.

2. Naturally produced air pollutions - In this category come pollen, volcanic gases, marsh gas, etc.

3. Particulate matter - The come mostly from industries, power plants and refuse disposal. Most of this particulate matter has not yet been adequately identified.

4. Hydrocarbons - They are mostly discharged by motor vehicles and industries.

5. Ozone - On account of human activities, its level may rise in atmosphere.

6. Nitrogen oxides - Nitrogen oxides include mainly nitrogen monoxide, nitrogen oxide, nitric acid mostly released by automobiles, power plants and industries.

7. Sulphur compounds - These Compounds include SO2 H2S and sulphuric acid. They are mostly released by fossil fuel i.e. coal, etc. based power generating plants and industrial units as refineries.

8. Carbon compounds - Carbon compounds include carbon dioxide and carbon monoxide. Carbon dioxide is released by complete combustion of fossils fuels. The

carbon monoxide is released by automobile exhausts.

Effect of Air Pollution- Man and Environment

Air pollutants can seriously harm materials, crops and livestock. They can spoil fabrics, surface coatings, metals and other commercial materials:-

1. Fluorides and arsenic found in air have caused serious injuries to livestock and cattle. When eaten, this fodder possess the cattle.

2. Ozone produces damaging effects on fabrics, dyes and rubber.

3. It is detected that Sulphur compounds can corrode or damage many metals, stones, electrical equipment, painted surface, textiles, paper and leather.

In fact, most of the air pollutants can cause human sickness or even death. Cases of air pollution disasters have been recorded from many countries. In the town of Donora, Pennsylvania, in October 1948; a temperature inversion trapped the emissions of certain industrial establishments and factories. Nearly 5,000 inhabitants fell ill, and at least 20 of them died following this calamity. A 50 day episode of severe air pollution in London in 1952 took a toll of some 4,000 deaths.

The researches done by Hickeyt and associates in 38 large American cities have highlighted that increasing concentrations of nitrogen dioxide and SO2 in the air are statically correlated with increasing death rates from chronic diseases associated with ageing and senescence, e.g. cancer, heart diseases, diabetes mellitus, etc. The air pollutants can and do cause long-term, slow and chronic effects on health in addition to their immediate and acute effects.

Some studies have indicated that the atmospheric SO2, NO2 and lead can cause mutations in human beings. These noxious gases are ocular and respiratory irritants. Concentrations of SO2 and No2 in the atmosphere constitute statistically significant predictors for maritally rates for various diseases of aging, including cancer and arteriosclerosis. In the same way, the lead concentrations were significant predictors for infant mortality rates and congenital malformations. Atmospheric NO2 can cause the potent mutagen nitrous acid and possibly bisulfites atmospheric chemical mutagens alter resting DNA in such a way that mutations appear after subsequent replications. Replication errors could sometime lead to a gradual accumulation of nutational errors.

Polluted air of big cities also contains carcinogens, e.g., benzopyrene, and other polycyclic compounds. Main sources are tobacco smoke, automobile exhaust and effluents from industry and electric power stations. The various kinds of particulates in the atmosphere cause soiling of various objects and are responsible for various nuisance effects. High dust levels in some occupational environments are also definitely harmful regardless of chemical composition.

Evil Effect of Air Pollution On Environment

1. Historical Monuments - A expert Committee appointed by the Government of India on Environmental Impact of Mathura Refinery has expressed the view that there may be a serious threat to the Taj, and that appropriate precautions should be taken to preserve the Taj from the adverse effects. They have warned that the Taj and various other ancient monuments may be irreparably affected by the corrosive fumes and pollutants discharged from the refinery, so much so that more money have to be spent on protecting Taj Mahal from the adverse impact of the Refinery than it took Shah

Jhan to buiid it. The work done at the National Environmental Engineering Research Institute at Nagpur has revealed that SO2 and other emissions from refinery and foundries can certainly affect marble as well as the red sanstone buildings of the tomb of Ehmad-ud-dulah, the Fatehpur Sikri complex, and the Red Fort. Some of the ancient temples in Mathura and even the famous bird sanctuary in Ghana forest near Bharatpur may be affected sooner or later.

This is the reason why the Department of Environment, Government of India, has ordered the closure of two coal-fired power stations in Agra and has also persuaded the railway authorities in Agra to switch over from coal to diesel. With a view to protect the Taj Mahal from pollution effects. A green belt is also being created around the Taj for the same purpose.

The crude oil to be processed in the refinery contains up to about 2 percent sulphur. This is oxidized to SO2 and will be emitted from the chimney of the refinery. Thereafter, in the moist atmosphere it is converted to dilute acid which causes decay of stone and marble.

2. Effects on Plants and Animals - Air pollution affects crop yields. Studies carried out in England have revealed great reductions in they yields of such plants as radish, lettuce, various flowering annuals and other crops when grown in the more highly polluted parts of big industrial cities. The smelting of ores causes serious damage to vegetation. Pollutant particles some time settle down on levels or nearby trees and may block the stomatal pores. In this way they affect gas exchanges, water relations and other physiological processes. Ozone can cause damage to palisade cells which lose water and disintegrate. At levels of 0.02 ppm, it damages tobacco, tomato, bean, pine and other plants.

Plants possess a gaseous environment. When the composition of this environment exceeds the capacity of plants to tolerate or adapt to the change, some stress is imposed. Thus the sensitive components of the ecosystem start malfunctioning. When the gas reaches inside the leaf, the cell membranes are affected. Alterations take place in SH/SS ratio. There can be permeability dysfunction as well. Free radicals are formed and ion transport can be inhibited and normal functions disrupted.

Plants growing in polluted environment often show symptoms of injury, and general debility. In temperate counties some findings are as under:

a. Reduction in fruit size.

b. Necrosis of fruit tips.

c. Necrosis of stigmatic surface.

d. Hardening of floral buds.

e. Tobacco to ozone and ptunia to photochemical smog. In this topics, mango is especially sensitive to polluted air. The aerial parts of the trees growing near thermal power stations, brickkilns, steel factories and smelters exhibited.

3. Cloud Structure - The particles called ice nuclei found in our atmosphere in low concentrations modify the cloud structure at subzero temperatures and the effects of ice nuclei are some time so severe that the precipitation potential of such cold could may also become altered. Two main sources emanating ice nuclei are steel miils and automobile emissions.

Forest fires and pulp and paper mills are the sources of certain other particles called could condensation nuclei. The CCN particles can modify precipitation from clouds that have temperatures above 00C. Other air pollutants like acid fumes, aerosols, ammonia, and SO2 affect the pH of rainwater. Certain chemicals in the discharge can interact in the atmosphere resulting in an amplification of their effects e.g., although SO2 and NH3 by themselves react slowly, their reaction rate is greatly increased in the presence of could droplets.

Surface active pollutant molecules are able to coat cloud particles and influence could and aerosol coagulation and evaporation. Some air pollutants are able to affect aerosols involved in could processes. The structure and distribution of clouds can be effected by such pollutants. It causes alterations in precipitation and optimal scattering properties. Besides, affecting some aspects of weather, these pollutants are also now believed scattering properties.

4. Acid Rain - The acid rain causes both air and water pollution. Some scholars have tried to identify the problem of acid precipitation which causes both air and water pollution. Some have even gone to the extent of describing acid precipitation as one of the gravest threats man.

To understand the concept of acidity is a basic requirement in any attempts to design ecosystem studies for answering questions about. The classical definition describes acids as proton donors and bases as proton acceptors. This means that acidity is a relative property and a substance may simultaneously act both as an acid and a base.

Acid rain is the principal cause of the acidification of soils and fresh-waters in several countries. It may b e deemed as rain with pH of less than 5.6 It follows from the definition of acidity in natural waters. The pure water in equilibrium with atmospheric CO2 has pH of 5.6 at 1000 C and zero alkalinity by definition. Thus rain with pH less than 5.6 is acid rain. The input of acid from the atmosphere imposes and acid load on ecosystems which receive it.

It is discovered that the HCI in acid rain comes mostly from the combustion and decomposition of organochlorine compounds and polyvinyl chloride plastics. The sulphuric acid comes from oxidation of sulphur in fossil fuels, and nitric acid comes from nitrogen fixation.

Acid precipitation affects water bodies. Lakes can lose as much as 70 percent of their bicarbonate alkalinity and in decade, due to the input of acid precipitations. Such a loss a alkalinity makes it difficult for crustaceans to moult their exoskeletons; in some cases, many crustaceans may die. Another indirect effect of the fall in water pH is the release of heavy metals such as mercury and aluminium in water, these metals in turn produce adverse effects on the indigenous biota.

5. Ozone Formation - Ozone, like several other photochemical oxidants is formed primarily from nitrogen dioxzide. Ozone like other oxidants such as PAN, harms plants. Ozone can also react with nitrogen monoxide, whereby the ozone is used up and nitrogen dioxide forms anew. If only these compounds are present, the ozone levels will not be very high. But when hydrocarbons enter into the picture, really high levels of ozone are reached. The reason is the hydrocarbons can transforms nitrogen monoxide to nitrogen dioxide without using up the ozone.

6. Photochemical smog - Photochemical smog is caused, in the presence of sunlight, by hydrocarbons and other organic substances and nitrogen oxides, most of which arise from the fumes and exhaust of motor vehicles. Smog formation atmosphere, the chief reaction

being the formation of certain free radicals of oxygen, hydrocarbon and nitrogen dioxide.

In Photochemical smog are found mixtures of ozone, nitrogen dioxide, nitrates, and other gases or dry particles. Intermittent fog in the same geographical area can be acidic with high concentrations of nitric acid. The various components of photochemical smog are toxic to agricultural crops and forest tress. Photochemical smog is known to cause serious health hazards. A particularly serious episode of heavy smog formation occured in July 1970 in Tokyo, New York, Rome and Sydney. Epidemics of such diseases as asthma and bronchitis occurred immediately after the smog.

Measure To Control The Air Pollution

Today it has become possible to control several types of air pollutions by employing modern technology and scientific researchers. It involves the benefits to improve the quality of the environment and human health, agriculture and vegetations. It also improves the ecological balance and stability of ecosystem. Some suggestions are as under:-

1. Control Through Law - Acts of Industries should be enforced to prevent gaseous pollution. Both the air pollution and noise pollution are included under Air and Noise Pollution Prevention Act. Legal steps already been taken by state and central government, to control the air pollution. Education can also play a vital role in preventing and controlling air pollution. It is essential to arouse wide awareness and to develop consciousness among students about air pollution and its resources and ways. Formal, non-formal and adult education programmes should aim at arousing and developing feelings and attitudes or consciousness about the problem which has serious adverse effects on human

health. Dr. Vibasi has suggested for preparation of Charts and mottoes on air pollutions, resources and remedial measures. Media and newspaper slogans should be used Save the Oil protect the air pollution.

2 Control on Industrial Air Pollution - The following methods are suggested to be used to control air pollution from gaseous pollutants -

a. Wet Dry Systems - This method is found successful in dry cleaning plants, printing shops, paint factories, food processing plants, etc.

b. Dry System - When water comes in contact with sulphur dioxide, it produces sulphuric acid.

c. Wet system - The alkali liquid reacts with sulphur dioxide to produce a precipitate.

3. Control on Automobiles Air Pollution - Some measures to reduce air pollutants from the transport sector in the advanced countries are as under:

a. To control through law such as Motor Vehicles Act, 1988 and other Acts of engine.

b. To use the filters.

c. Control evaporation from fuel tank and carburettor.

d. To check pollutants emission from vehicular lexhaust.

e. To adopt speed limits for road traffic at a maximum of 100 kph.

f. To adopt properly regulated emission standards for all new motor vehicles, including aeroplanes and ships.

g. To adopt plans and timetables with a view to reduce significantly the total European volume of road and air traffic.

h. To implement volume control, no investment in

public transport and railway freight systems should be increased.

4. Various Measures - A range of measures should be adopted including as under :

a. Applying best available techniques in the energy, transport, industry, and agriculture sectors.

b. Switching to less polluting fuels.

c. Increasing the efficiency of energy production, transmission, and use.

d. Increasing to conservation of energy.

e. Using renewable energy sources.

5. Ascertainment of Critical loads - With a view to check the continuing deterioration of the environment, concentrations and depositions of air pollutants must be reduced to below the critical loads. We can define critical loads as a quantitative estimate of an exposure to one or more pollutants above which adverse effects on receptors, such as plants, ecosystems or materials, may take place.

Soil Pollution

We can define soil pollution as the contamination of soil with excess of chemicals, fertilizers, insecticides and herbicides. Soil is very important environmental component for human beings, animals and plants. It is the basic medium for food, vegetation and natural resources as well as the base for the development of human culture and civilization. Various aspects and components of soils and texture, structure, moisture, profiles and horizons.

The decrease in the quality of soil either due to human activities or natural sources or by both is give the name soil pollution or soil degradation. It is caused by various reasons as under:

1. Lack of Human content.
2. High fluctuations of temperature.
3. Excess of deficit of moisture content.
4. Decrease in soil micro organisms.
5. Decrease in plant nutrients.
6. Soil reason.

Main Causes of Soil Pollution

There are various causes of soil pollution. Accelerated rate of soil erosion, silk eruption, deforestation, excessive use of pesticides, insecticide, chemical fertilizers and herbicides, polluted waste water from industries, forest fires, urban areas a few micro-organisms, dummying of urban and industrial wastes, water-logging and related capillary process, drought, leashing processes etc. Main causes of soil pollution are as under:-

1. Resistant Object and Recalcitrant Wastes - Various large, resistant objects such as cans, tyres, other metallic or plastic junk or even parts of old cars, etc., act as pollutants. The polychlorinated biphenyls is a serious pollutant. At present PCBs are used primarily as dielectric fluids and coolants for electrical equipment. They are also used in paints, paper coating, carbon-less paper and chlorinated rubber. Mineral oils can become contaminated with low concentrations of PCBs. It should be noticed that PCBs are highly stable under most conditions and this resistance to degradation increases their polugenic nature.

2. Metals - Metalliferous mining, smelting, and refining release metals to the soil atmosphere, water and dediments, e.g. electricity generating plants, incinerators, sewage sludge, and residues from leaded gasoline. Studies are made to know the long-term effects

of metal accumulation in the soil rustling from applications of sewage sludge.

McGrath reports the results of a long-term experiment which received metal-contaminated sewage sludge from 1942 to 1961 and inorganic fertilizers thereafter. Additions of sewage sludge and a sludge-straw compost increased considerably the concentration of Zn, Cu, Ni, Cd, Cr and Pb in the plough-layer soil, these have persisted for over 4 decades now. Metal concentrations in the soil started failing since 1941 when the sludge applications were stopped. Most workers agree that the soil and organic litter layer components of woodland ecosystems are the ultimate sinks for heavy metals in aerially contaminated areas. It is found that once soils have become contaminated, they are unlikely to be depleted of lead and other metals by leaching.

3. Fertilizers and Detergents - Some of fertilizers and detergents are fairly a very serious pollutant. These particulates originate from the weathering and erosion of rocks. Silt particles are mainly composed of silicates and carbonates with some phosphates and heavy metals absorbed on them. Silting can choke irrigation channels and rivers. Edible fishes, e.g., salmon, are unable to spawn on silted gravel beds.

4. Silt - The dirt and dust particles which are carried from land to water. It constitutes a very serious pollutant. These particulates originate from the weathering and erosion of rocks. Silt particles are mainly composed of silicates and carbonates with some phosphates and heavy metals absorb on them. Silting can choke irrigation channels and rivers. Edible fishes, e.g., salmon, are unable to spawn on silted gravel beds.

5. Weedicides - Residues or various weedicides and insecticides often accumulate in agricultural soils

rapidly. Insecticides are designed to kill insects. Some herbicides differ from insecticides in killing both desirable species as well as the intended target. Their impact may not be confined to the plant community or agricultural crop but they may also act indirectly through effects on such soil microbes as nitrogen fixing blue-green algae and bacteria. This is turn may cause harm to the growth and production of higher plants. The impact of such chemical wastes as phenols, acids, metals, etc. applied to soil can also be fairly high and may tend to complete destruction of vegetation and also soil sterilization.

6. Man's Activities - Almost of man's activities produce wastes. Many wastes can be recycled or reduced at source. Yet there are but some wastes and their residues which are inevitable and have to be placed in landfill sites. U to 80-90 percent of our refuse has generally been disposed off on the land. Hazardous wastes come from industries, agriculture, and other sources. In fact, much of a nation's prosperity and economic well-being originate from activities that have produced hazardous wastes. Their proper disposal often presents tricky problems and needs careful consideration. However, some big chemical companies are now installing huge incinerators with a view to handle the wastes produced by them.

7. Acid Mine Drainage - Acid main drainage is chief source of pollutant. The reason is acid generation and discharges continue even after mining has ceased. Coal mining operation releases some acid arising from the exposure of iron sulphide minerals found in the coal for the purpose of keeping control, certain steps may be followed during the mining operations to prevent the acid from being formed.

8. Mining Activities - Mining activities constitute one of the largest producers of solid waste. Extraction of fuel being a destructive process, environmental degradation is an inevitable result of such mining. The various environmental pollution problems arise from mining. The chief is of solid wastes. The handling and disposal of the solid waste is the principal cause for disturbed lands. During mining and benefaction, environmental problems can arise from the runoff from the disturbed area, and dust from blasting, extraction, haulage, storage, and grinding, etc.

9. Municipal Solid Waste - It is reported that most municipalities face the problem of managing the ever-increasing volume of municipal solid waste in air environmentally acceptable manner. An integrated strategy using the Four Rs. is regarded as the most efficient and cost effective way reducing the MSW requiring landfull, yet it is not being done. So there is continued reliance on landifilling. Given the current landfill capacity crisis, incineration may gain support as an integral part of an effective waste management strategy.

Effect of Soil Pollution

Main effects are as under:-

1. Evil Effects of Fertilizers on Soil - Some adverse effects of fertilizer on soil use are:-

(i) Disturbance in the ionic balance and equilibrium in soils. High acidity, nutritional imbalance, shortages of certain trace elements and molydbenum or selenium toxity.

(ii) Stimulating of weed growth in crop fields.

(iii) Deterioration of water resources caused by eutrophication.

(iv) Changes in mobility status of various nutrients in soils.

2. Evil Effect of Nitrates - The excessive and indiscriminate application of inorganic fertilizers is found leading to accumulation of nitrates in water. When such waters are drunk by cattle or human beings, these nitrates are reduced to the toxic nitrites by intestinal bacteria. Nitrates can cause a serious diseases known as methemoglobinemia damaging respiratory and vascular systems and may even cause suffocation.

3. Evil Effect of Asbestos - Asbestos is widely used as in insulating material. Its recognition as a pollutant was realized in 1973, when the American Environmental Protection Agency reported that the water of Lake Superior contained hazardous amounts of asbestos as the iron ore wastes were being dumped in the north western portions of the lake. The recent medical opinion attributes such hazards as cancer of lungs, stomach and intestines to asbestos pollution. Asbestos is now also recognized as a dangerous air pollutant since its particles can be carried into the atmosphere.

4. Effect of Pesticides - The pesticides of their residues pollute the soil as under:-

(i) They may cause the adverse effects as resistance, fungal displacement and other population changes.

(ii) Many of them persist and thus cannot be disposed off.

(iii) Their concentration and magnification in biological systems may lead to certain unexpected or untoward results.

(iv) They strike the intended pests as well as several non target organisms.

3. Health Hazards - Pesticides pollute the

environment causing health hazards. A decade ago, a mysterious disease involving pain in joints afflicted poor farmers in several villages in Karnataka State. It was detected that the crustaceans had become contaminated with parathion and endrin which were used in the paddy fields. The pesticides ultimately were ingested by the farmers. As such, they developed the disease.

4. Poverty and Malnutrition - In developing countries, poverty, malnutrition and weak health of the agricultural workers can be linked with improper storage of the chemicals and the use of pesticide containers for storing food and drinks in villages. These factors contribute to accidents of pesticide poisoning. It is observed that most poor peasants live in a single-room hutment where the probability of pesticides getting into the food which is often cooked in the same room is high. Besides, the access of the chemicals to children is easy. These hutments are frequently very close to the crop fields. When the crops are sprayed, the spray frequently drifts into the house, exposing the children to grave risks.

5. Effect on Insects - The effects of pesticides on various kinds of insects , earthworms, invertebrates, protozoa, and microbes found in soils, especially the reduction in natural enemies due to pesticide usage, increased resistance to pesticides in insects and mites, and honey bee poisonings account for about 70 percent of the calculated socio-environmental costs for pesticides in the USA.

6. Effect of Honey Bees - Bees aid the pollination of several plants. Pesticides have significantly affected some honey bees and other useful insects so their populations have declined. According to Pimental et al. annual agricultural loses due to poor pollination from pesticides can be as high as $ 4000 million in the USA.

Some effects are as under:-

(i) Drifting or leaching pesticides drain into nearby water in rotation. Drifting or leaching pesticides drain into nearby water bodies, and cause fishery and wildlife losses. It is observed that wild birds and mammals exposed to pesticides suffer by death from direct exposure to high doses.

(ii) Persistent herbicides also can injure crops planted in rotation.

(iii) Herbicides that drift from a treated crop to a nearby crop also cause serious environmental problems.

(iv) Crops are damaged instead of being protected by improper pesticide applications e.g., application of improper dosage under unfavourable conditions.

7. Effect on Ladybird Beetle - The consequences of pesticides are almost invariably adverse and harmful. Here one case study can be cited. In the nineteenth century, the ladybird beetle was brought from Australia to California to control a scale insect pest of oranges. The beetle kept the pest under check for more than five decades. However, in 1946 when DDT began to be used in the citrus orchards, its existence was endangered. The beetle was susceptible to DDT and hence its population declined. This led to a marked increase in the population densities of the pest. A subsequent withdrawal of DDT again restored the natural balance of biological control within a few years. It is established that DDT is one of the most effective pesticides known. It was banned in the USA in 1972.

It is believed that the most serious threat to public health is from the uncontrolled use of pesticides for agronomic practice. Chapping and Wessertrom found so in 1981. In India, widespread resistance to pesticides has developed in Anopheles spa particularly in areas which have recently shifted to high-yielding farmers of agricultural production Resistance may be clearly seen in the areas where Green Revolution technology, has been introduced.

8. Effect of DDT - Extensive researches in the USA have established widespread distribution of DDT residues through food chains in several lakes. Residues were detected in shallow and deep water mud samples, the whitefish, duck, crustaceans ring-billed and herring gulls and other fauna adversely affected. Residues of Dieldrin, Endrin, DDT and DDE have been detected in all major river basins in the United States. It is also believed that the declining percentage of immature birds in the American bald eagle population is due to DDT and other hard biocides.

Control Measures

1. Processing Sludge Ash - It is proved that processing sludge ash into lightweight aggregates for the concrete industry is the most promising utilization option. Lightweight aggregate may be produce by firing pelletized or slabbed sludge ash at 1100 to 12000C produce a hard, ceramic-like material. It can later be crushed, sized and graded to conform to required aggregate specifications. The low thermal conductivity and high fire resistance of the aggregates make them suitable for thermal conductivity and high fire resistance of the aggregates make them suitable for use in thermal insulation and fire protection applications such as fireproof wallboard.

2. Ban of Pesticides - As the surviving biocide residues are found uncontrollable, insidious and unpredictable in their actions, Rudd has advocated a complete ban on such pesticides whose toxic effects

cannot be confined to the site of application. In fact, substantial harm has already been done by these biocides. In such a position even if their future use were banned now, sufficient amounts of residues will still persist in the various ecosystems to cause serious ecological effects for decades.

3. Technologies - Attempts are under way in Canada and other countries to evolve suitable technologies for destroying PC Bs contained in both liquid and solid wastes.

4. Limitations on Concentrations of Harmful Substances - Some limits are imposed on the concentrations of harmful substances in the effluent, however not on the production rate of the mine. The maximum authorized monthly average concentration prescribed in Canada is 0.5 mg/litre for arsenic, nickel and zinc; 0.3 mg/1 for copper, 0.2mg 1 for lead, and 10 pCi/ litre for randium 226.

5. Limitation By Enactments - In several countries, regulations are enacted to limit the concentrations of the following metals in the mining effluents: copper, lead, arsenic, nickel, zinc and radium 226.

Here mines are taken as all metal mining and milling facilities and associated smelters, refineries, acid plants, pelletizing plants, sinter plants and any similar operations where the effluent from several such operations is combined with effluents form mining and milling.

Noise Pollution

The word noise is an unwanted or unpleasant sound that causes discomfort. Noise is Wrong sound, in the wrong place at the wrong time. Noise pollution is the unwanted sound dumped into the atmosphere leading and health hazards. Noise Pollution may be defined as the state of discomfort and restlessness that is cause to man by unwanted high intensity sound known as noise. Unbearable high pitched sound is called noise pollution.

Effects of Noise Pollution

Some effects of noise pollution are as under -

1. Psychological Effects - The noise pollution exercises psychological effects. It causes a great number of behavioural change among human beings as well as animals. Disco music some times, causes temporary deafness in the listens.

2. Non-auditory Effects - These include the interference with speech communication, physiological disorders irritation/annoyance and loss of working efficiency etc.

3. Auditory Effects - As clear from the name, these include auditory fatigue and deafness. Deafness is caused by continuous noise exposure in the biggest cities like Delhi, Mumbai, Kolkata and Chennai, etc.

Auditory fatigue appears in the 90 dB and may be associated with side effects as whistling and buzzing in ears. Deafness can be cause due to continuous noise exposure. Temporary deafness occurs at 4000-6000 hz. Permanent loss of hearing occurs at 100 dB. Mumbai and Kolkata are the nosiest cities in the world.

Methods of Control Noise Pollution

Main methods to control noise pollution are as under-

1. Education - The students and public should be educated about the harms caused by noise nuisance though news media.

2. Preventive Acts or Law - Air pollution Control Act is enacted to check the noise pollution. In should be enforced strictly.

3. Plantation or Vegetation - Big trees should be planted along high ways, streets and other noisy places. They absorb noise.

4. Ear Protection - Various devices such as ear plugs and ear muffs may be provided the workers of industries.

5. Transmission Control - Another suggestion is to cover the room at source.

6. At Source Control - Another suggestion is to cover the room walls with should absorbers.

Marine Pollution

Our age long concept that oceans are limitless or bottomless is erroneous for the oceans are quite limited in respect of their capacity to act as huge sink and receptacles for wastes.

Ocean pollution can be cause by ship-generated discharges of oil and petroleum products, noxious liquids, packaged dangerous goods, sewage and garbage, etc.

Process

The fate of pollutant discharged into the sea is determined by many route and processes. It is studied that the pollutant may become dispersed in any of the following ways:

1. Such biological processes as may include uptake by seaweeds, fish, birds and phytolankton. From the latter it may pass on through a food web involving zooplankton, fish and mammals.

2. Transported by cocoon currents and migratory birds.

3. Turbulence and ocean currents.

Sources

Terrestrially produces or eroded substances and

pollutants are mostly carried into an ocean through:

1. Deliberate disposal of substances by man through pipelines.

2. Rivers.

3. The atmosphere.

In fact, oceans can cope up with reasonable amount of particulate minerals, common soluble chemicals, mud and salt, but they are unable to rot or decompose organic matter. Organic materials however, are generally:

(i) In sedimates containing high concentrations of fine particles of Ca Co3 precipitation of disclosed phosphate occurs on the surfaces of CaCo3 grains. As such, phosphate concentrations in the interstitial water of such sediments much lower than those in anoxic siliceous sediments.

(ii) Heavily polluted sediments are subject to faster decomposition of their organic matter than less polluted ones.

(iii) Dissolved ammoniac and phosphate originate from the same general organic decomposition process which in some marine sediments is believed to be brought about by microbes in addition to those concerned with sulphate reduction.

(iv) Eaten, digested and execrated by marine animals.

Effect of Sea Pollution

1. It is revealed by the studies of fishes in polluted and non-polluted areas near Helsinki that the production and biomass of fish in polluted areas was great yet the species diversity was less.

2. The greatest diversity of bottom fauna was found in clear or only slightly polluted water and it was found to decrease progressively with increasing load of

pollution.

3. Most crustaceans avoid polluted areas.

4. The most common pollution-tolerant benthic animal of Macoma ballica.

5. In highly polluted areas, no macroscopic bentic animals are found however chironomid larvae occur at the sea bottom.

6. The bottom fauna of polluted coast shows a zonation in keeping with the extent of pollution.

7. Any serious pollution of seawater can result in a loss of diversity in the marine organisms and unstabilize and simplify the otherwise complex and stable ecosystem. The pollution is the Baltic Sea as reported by Bagge, along the coast of Finland mainly comes from sewage and effluents from wood-processing industries.

Oil Pollution

In recent years oil pollution has become formidable. Recent increases in shipping and drilling operations have caused increased contamination of our oceans with chemical pollutants, including petroleum hydrocarbons. These adverse effects are as under:-

1. Aquatic species may be more sensitive than mammals to the acute lethal effects of a variety of chemicals.

2. Marine fish, crustaceans, molluscans, and zooplankton accumulate hydrocarbons from polluted water.

Spills

Much of the oil and petroleum products are shipped by sea and any spill of this oil can cause serious pollution problems. Spills are occasional discharges or dumping that happen due to:

1. Convenience dumping into sewers, streams,

estuaries, coastal waters.

2. Deliberate dischages of ballast water.

3. Malfunctioning of vehicles and equipment.

4. Human error.

5. Accidents.

Some of these spills cause obvious, damage to marine fish, birds, invertebrates, algae, and mammals. Sometimes the spills themselves may not be very harmful to marine life, but the detergents used to clean up the spill prove extremely harmful. It is studied that those areas of British coasts around which large amounts of toxic detergents were used to clean up the oil slick caused by the notorious Torrey Canyon accident a few years ago, witnessed a high morality of marine life.

Thermal Pollution

Thermal pollution is the pollution caused by the discharge of heat in the form of hot water, air of effluent into the environment. Many industries use a log of water for cooling purposes and return this water to the stream at a higher temperature. High temperature can exert a lethal effect on the biotic communities in an equated habitat. Some animals are killed our-right by the hottest water, with less hot water some harmful effects can result e.g. a bacterial disease of salmon in some rivers is though to become aggravated by a slight rise in water temperature.

Effect of Thermal Pollution

The thermal pollution can affect the entire ecological balance of stream. For Instance:

1. In USA Warinner and Brehmer found that carbon assimilation by natural phytoplanktons was affected by an artificially induced increase in water temperature.

It is found that the greater the temperature rose caused by thermal pollution, the greater was the depression in primary production.

2. Many research workers have narrated their findings regarding the effects of thermal pollution on microbes, plankton, plants and animals.

3. Discharge of waste heat in rivers can not only affect the indigenous fish population but may also affect their migration or spawning.

4. The rapidly spreading weed Elodea condenses has become the dominant species at thermal outfalls and heated areas in Alberta and that is such areas this weed has tended to replace species of Myriophyllum, Chara and Potamgeton.

5. It affects species composition and production of submerged macrophytes.

6. Heated water produces more serious effects in those cases where an important biological resources in diminished. The sessile, benthic invertebrates cannot move freely so. Hence in their case a plume can exert a much more profound effect.

7. High water temperature adversely affects the daily and seasonal behaviour and metabolic responses.

8. Salmon often fails to spawn in such conditions. Warm water holds less oxygen and hence its BOD rises. Water 0oC has approximately 14 ppm of dissolved oxygen whereas that at 200 has only 9 ppm.

9. The trout eggs commonly fail to batch in water that is too warm.

10. The discharge of warm water may cause fish eggs to hatch so early in the spring that the natural food organism needed by the young fingerlings would not be available.

7

HEREDITY AND ENVIRONMENT

Far from satisfying our psychological curiosity, the fact of individual differences in intelligence and personality spurs us on to trace out the causes of these differences. The question thus raised is a large scientific problem with important practical bearings. For if it is true, as some believe, that one person's superiority over another due mostly to heredity, then the best hope for the future of mankind lies in improving the breed. If the contrary view is nearer the truth, and the main causes of individual differences lie in the environment, then it is the environment rather than the breed which needs to be improved. Shall the gardener pin his hope on careful cultivation of the soil, or on selection of the best seed? The practical gardener knows perfectly well that both are necessary.

He could never grow a superior crop from inferior seed, no matter how rich the soil, nor from even the best of seed down in poor soil. The development of a plant is clearly a response of the heredity present in the seed to the environmental stimuli of soil, moisture and sunlight. It is equally true of the animal that development depends on both heredity and environment, and that one individual differs from another sometimes because of different heredity, sometimes because of different environment, and usually no doubt from both causes combined. This rather dogmatic statement needs to be scrutinized in the light of the main facts in individual development, and it is not enough if we push our study back to the very birth of the individual, for both heredity and environment have exerted decisive influences during the prenatal period. We need to see the young individual's mental development against the background of early physical development against the background of early physical development as studied by the biological sciences of genetics and embryology.

EXPERIMENTS ON HEREDITY AND ENVIRONMENT

The logic of investigation is clear from what has just been said. To bring out the importance of heredity, take two individuals of different heredity and let them develop in identical environments; if they develop differently the cause must certainly be the difference in heredity. Similarly, in investigating the influence of environment, the experimenter takes individuals of identical heredity and lets them develop under different environmental conditions, so that any difference which results must be attributed to environment. These two types of experiments have been tried many times and in many ways on plants and animals.

Monsters prove the importance of environment in causing individual differences

Is development what its name would seem to imply, a mere "unfolding" of characteristics which are predetermined by heredity? The negative is proved by a second type of experiment, in which an individual of normal heredity is subjected during early development to abnormal environmental conditions. There are obvious difficulties in the way of performing this type of experiment on unborn mammals, but the eggs of birds can be tampered with during incubation, and those of fishes and amphibians as they develop in the water can be heated or chilled or treated with chemicals or electric currents or rays of any kind. The result is frequently the production of a "monster," an individual differing in some striking way from the norm of the species. For example, a collection of young fish embryos, just as the stage when the eyes begin to form, is placed in the icebox for a number of hours, with the result that some of the fish become Cyclopen monsters, having a single centrally located eye.

The difference between these monsters and normal two-eyed fish is due to a difference in environment. Instead of saying that heredity unfolds itself, it would be nearer the truth to say that development consists in the unfolding of heredity by environment. Heredity is like a rocket lying on the shelf, a potential firework, containing within itself some of the conditions for a display, but dependent also upon external conditions. One display may differ from another be cause of a difference in the way the rockets are made or because of a difference in the way they are set off.

Hybrids prove the importance of heredity in individual differences

If a male of one species is mated with a female of a different species, the offspring is known as a hybrid. With the male parent a donkey and the female a horse the hybrid offspring has the same prenatal environment as a normal horse and should therefore be born a horse, if only environment counts. But the new born hybrid turns out to be a mule, differing considerably from a horse. Moreover, though the postnatal environment may be the same for both, the young mule becomes more and more a mule, and the young horse more and more a horse.

The difference between a mule and a horse must therefore be laid to heredity. Half of the mule's chromosomes, in fact, are donkey chromosomes and are responsible for his unhorselike peculiarities. Cross-breeding experiments in great variety have been made by students of genetics, and the wonderful body of knowledge thus achieved from the day of Gregor Mendel, 1866, down and especially during the present century should certainly be examined by anyone who is skeptically inclined towards heredity.

ORIGIN AND EARLY DEVELOPMENT OF THE INDIVIDUAL

About nine months before birth the human individual begins his career, in the from of a one-called organism, the fertilized ovum, which is an ovum from the mother's gonads combined with a spermatozoon from the father. In the favourable environment supplied by the mother, this microscopic individual grows and divides into two cells, into four, into eight, into many, many millions. Along with growth occurs 'differentiation,' which means that the cells derived from the fertilized ovum, the 'daughter cells,' differ among themselves, with the result that the individual develops a great variety of tissues and organs. Amid all this multiplication and differentiation, the parts of the organism remain in union and interaction with each other, and the individual remains an integral whole, a perfect example of unity in multiplicity.

Environmental factors in development

A hen's egg, though fertilized and containing all the

heredity necessary for the development of a chick, remains quiescent necessary for the development of a chick, remains quiescent while in cold storage and develops only when incubated. Environmental conditions must be favourable or the fertilized ovum will not develop at all, and if they are unusual development is apt to take an unusual course. The prime conditions to be met by the environment are protection, warmth, food, oxygen and the elimination of waster products.

It might seem, and indeed has sometimes been asserted, that heredity, however potent in the first stage of development, must become less and less important as the effects of environment accumulate. If all of heredity is present in that microscopic cell, the fertilized ovum, how can it be more than an infinitesimal factor in the individual at a later stage, as at birth, when eight pounds of material from the environment have been incorporated? The answer is that this absorbed material has not simply been added: it has been built into the organism in accordance with the organism's own hereditary characteristics. All the milk absorbed by the infant in his first year does not make him more and more milky, for he changes it chemically into human muscle, bone and brain.

The chromosomes keep pace with the growth of the individual and are just as potent a factor in the later stages of development as at the very beginning.

No native traits, and yet native differences

What we have just said forces us to desist from speaking of native and acquired traits, "native" meaning due to heredity and "acquired" meaning due to environment. Every trait of an individual depends on both heredity and environment. This conclusion certainly seems at first thought to run counter to common sense. If one individual has blue eyes and another brown, is not the difference due to heredity and to heredity alone? If one speaks Swedish and the other Italian, is not this difference due entirely to the differing social environments in which they have grown up? Yes—but nothing we have said forbids us to speak of native differences and acquired differences. Consider the rectangle once more.

While the area of any single rectangle certainly depends on both base and altitude, the difference between two rectangles may lie wholly in one dimension. In the same way, two individuals of the same heredity may differ because of differing environments: and two individuals exposed to identical environments may differ because of differing heredity.

Chromosomes and genes

This bare sketch of embryonic development needs a little more details in order to reveal the effects of heredity and environment. Within the general fluid mass or cytoplasm of any living cell is contained a small nucleus of somewhat denser material; and while the cytoplasm performs the distinctive activity of the cell—the secretion of the gland cell, the contraction of the muscle cell, or the rapid conduction of the nerve cell—it is the nucleus that takes the lead in the processes of growth. In cell division the first thing that happens is a division of the nucleus.

The chromosomes rod-like bodies which make up the nucleus, deploy in regular order and split, each forming two daughter chromosomes one for each of the daughter cell. This splitting of the chromosomes occurs in every successive cell division, and thus the chromosomes of the fertilized ovum are reproduced in every cell of the developed organism. Every cell of the same individual possesses identical chromosomes. Now the chromosomes are-the carriers of heredity. Half of them were brought into the fertilized ovum by the spermatozoon and the other half were present in the ovum itself prior to fertilization. From these facts we can draw two highly important inferences: (1) the individual's heredity is derived in equal measure from the maternal and paternal sides: and (b) the individual's complete heredity is present in every cell of his body. The chromosomes are not ultimate units: each one is conceived to be an aggregation of numerous parts called "genes."

The genes are too small to be seen with the microscope, but the location of single genes within the chromosomes has been discovered by indirect means. Each gene carries a unit factor of heredity. The whole heredity of the individual consists of many unit factors corresponding to as many genes. That is to say, unit factors of heredity have been located in definite parts of certain chromosomes. The gene is of course a theoretical concept, like the molecule or atom or electron, with the same sort of scientific justification.

Development is the product, not the sum, of heredity and environment

The relation of heredity and environment is not like addition but more like multiplication. The individual does not = heredity + environment, but = heredity × environment. Let heredity be represented by the base of a rectangle, and environment by its altitude. Then development, or the developed individual, is represented by the area of the rectangle, the product of base and altitude. We cannot say that the area depends more on the base or more on the altitude; for if either should drop out there would be no area left. In the same way it is absurd to think of either heredity or environment as being more important than the other in the process of development. The conclusion that all development depends on both heredity and environment is of importance in psychology, since what holds good of development holds good of the developed structures and of the activities of those structures.

HOW UNIFORMITY OF BEHAVIOUR IS PRODUCED

In spite of all the differences which we have been discussing, is it not true that people are, after all, very much alike both in structure and in behaviour? They breathe alike, they pass through the same stages of growth, and within the same social group they talk and conduct themselves in much the same manner. Now if differences are due to either heredity or environment, to what are uniformities due? We easily fall into the error of attributing certain uniformities to heredity and others to environment. Uniformity in breathing is said to be handed down by heredity, and uniformity in language by tradition.

Biological uniformities are said to be transmitted by the chromosomes, and cultural uniformities by the social environment. But this way of thinking runs counter to our previous decision that all behaviour depends on both heredity and environment. Let us go back once more to our rectangle. Two or more rectangles cannot have the same area just because they are alike in one dimension; they must be alike in both, or else differences in one dimension must be compensated by reverse differences in the other. By the same logic, if two identical twins develop alike, they must have been exposed to practically equal environments.

Uniformity in development or behaviour means, not that either heredity or environment has been uniform for all the like individuals, but that both heredity and environment have been uniform, or else that the difference in heredity has been balanced by a reverse difference in environment. This is sound logic, but the conclusion does not appear entirely convincing. It would seem that individuals of differing heredity become more and more alike when kept together in the same environment. They speak the same language, no matter what language was spoken by their parents.

The reason is obvious: it is well within the native capacity of the average child to pick up the rudiments of any human language. The heredity of all normal individuals can be regarded as the same, so far as simple speech is concerned. Therefore, having sufficiently uniform heredity, they develop alike with respect to the rudiments of a language when exposed to the same environment. Our logic is not overthrown. A similar example is afforded by the use of the bow and arrow in nearly all tribes and peoples, the world over, prior to the introduction of firearms. This is called an instance of social transmission, but it depends also on uniform heredity to the extent of capacity for using the bow and arrow. None but human beings have the hereditary capacity for this mode of behaviour, and even the chimpanzees have not picked it up from their human neighbours.

SEX DIFFERENCES

Sex itself is determined by heredity. Of the 48 chromosomes of the human fertilized ovum, there is one which determines the sex of the individual. It occurs in two alternative forms, one giving male gonads, the other female gonads. Whatever differences between the sexes are dependent on the gonadal hormones are thus to be called hereditary sex differences. In addition, the sex chromosome contains other genes besides those which determine sex itself and these genes are sex-linked factors in the development of the individual, i.e., they are hereditary factors which differ from one sex to the other. We know little about them in the human species, but would expect them to amount to not over 1/48 of all hereditary factors.

These sex-linked genes, plus the gonadal hormones, are responsible for the familiar sex differences in the plumage of birds and in the size and build of men and women. Probably due to heredity is the well-known difference between men and women in muscular strength. Man surpasses not so much in endurance as in intensity of muscular action. He can squeeze harder and punch harder and run faster. This sex difference is not due simply to occupation, since it appears even in those tribes where the women do much of the hard labour. Men's muscles are somehow capable of using fuel and oxygen more rapidly, and their red blood corpuscles, which carry the oxygen from the lungs to the muscles, are ten percent more abundant those of women. Another sex difference comes to light in the earlier and more rapid maturing of girls.

They reach sex maturity and adult stature a couple of years younger than boys, and it is probable, though not proved as yet, that they reach mental maturity earlier. You have then a difference in size and in muscular strength favouring men and a difference in early maturity favouring women. Add to these the obvious difference in reproductive functions and you have perhaps listed all the important sex differences which are strictly hereditary. These few biological differences are responsible indirectly for considerable divergence between the occupations and achievements of men and women.

When we find among primitive tribes that man is the hunter and woman the gardener, we are inclined at first to say that it is just a matter of tradition; but we soon see that this division of labour is much more economical biologically than the reverse. If this case is typical, the behaviour differences which exist between men and women are neither forced by heredity nor due simply to tradition; for while the individual id following tradition, tradition is following lines of least biological resistance.

It is important to remember that sex differences in behaviour or achievement are differences between the male and the female averages, and that the distributions overlap. Even in stature, though men average two or three inches taller, many individual women are taller than the male average. The same overlap is found in muscular strength and is still greater in the more intellectual performances. Usually the sex difference in any performance in small in comparison with the range of variation within either sex.

Sex differences in test results

In intelligence tests of the Binet type, girls, show a slight superiority over boys of the same age, while in performance tests the slight advantage in with the boys. On the whole the sexes are practically equal in tested intelligence. Memory tests give the advantage to the girls. Their learning of either connected or disconnected material is more rapid than that of boys. Girls surpass in various language tests, such as naming all the words possible in three minutes, or saying the opposites of a given list of words. Girls seem indeed to have a definite advantage in all sorts of language activity. As babies they begin to talk a month earlier than boys, they pick up a vocabulary more rapidly, and they use longer sentences in the early years. In school they surpass the boys in language work. Boys surpass in the tests that have been prepared to measure general information; and in school they carry away more information from geography, history and the sciences. Boys surpass in most of the motor tests, though decided not in handwriting.

Their superiority in managing mechanical contrivances is as marked as that of the girls in language. Can we accept these as native sex differences? There is really no good evidence on the question. We can argue that boys simply have more experience with tools and mechanical toys, but the answer is that they certainly seem to have more spontaneous interest in such things. But possibly this difference in interest in merely the result of adult steering—possibly, not probably. We might expect to find greater differences between the sexes in personality than in intelligence, but so far as tests have gone there is not much that is definite to say. Would you expect men to be more inclined to introversion than women, or the reverse?

The best indications are that the sexes do not differ in this respect. On the whole the sexes are much more on a par mentally than our ancestors would have believed. Biology gives us no reason to expect large differences, tests do not reveal any. But we have to leave unanswered one specially intriguing question: why are the great geniuses and producers, and the great criminals and scoundrels, still predominantly men in spite of the increased freedom of women during the past hundred years? One theory is that the male sex is inherently more variable and so presents the more extreme deviations from the average, both upwards and downwards. Another possibility is that men have more energy because of their more abundant red blood corpuscles. Both theories are speculative at the present time.

HOW DIFFERENCES IN INTELLIGENCE ARE CAUSED

The human species presents a difficult field for the student of heredity and environment. The interval from one generation to the next is so long that investigation has to advance very slowly. Breeding cannot be controlled by the investigator who is therefore forced to deal with individuals of undefined heredity, and environment can be controlled only to a slight extent for scientific purposes. The logic of investigation remains the same as in the hybrid and monster experiments, but the difficulty is to find individuals of identical heredity or individual exposed to identical environments.

Identical twins are the only exception

It sometimes happens that two individuals come from a single fertilized ovum, by a splitting of the embryo at a very early stage of development. Theses two individuals therefore have identical chromosomes, identical heredity. At birth such individuals usually resemble each other so closely as to be scarcely distinguishable and this resemblance persists. It extends to such details as the finger prints. These are identical, or unioval twins. “Fraternal twins” occur when two separate ova are fertilized at practically the same time. Coming from different fertilized ova, with different chromosomes, fraternal twins are simply siblings, with no more in common, so far as heredity is concerned, than can be expected from two children of the same parents. Fraternal twins are as often of different sexes as of the same, while identicals are always of the same sex. Even when of the same sex, fraternal twins often differ considerably in appearance and constitution. Now the question is whether identical twins are more alike mentally than fraternals or than siblings. One convenient way of handling test data to obtain an answer to such a question is by use of the correlation coefficient.

The higher the correlation the closer the resemblance. Much work on anatomical measurements of siblings has indicated a normal correlation of approximately .50. In intelligence tests it is probably a little closer, though not above 60. Such moderate values of the correlation coefficient mean that siblings differ considerably. Between fraternals a little higher correlations are found than between siblings. Since fraternal twins are presumably no more alike in heredity than siblings, the higher correlations found suggest that the sum total of effective environmental influences, prenatal and postnatal combined, differs less between fraternal twins than it does between siblings born at different times.

When we come to the identicals, we find correlations

that are much closer than between fraternals or siblings. To explain this result in terms of environment we should assume that the environment is much more alike for two identicals than it is for two fraternals. This assumption appears rather strained, and the explanation in terms of heredity is simpler, since we already know the hereditary difference between fraternals or siblings to be much greater than between identicals. Another method of handling the data is simply to find the difference in IQ between the members of each pair of twins and so to obtain the average IQ difference between identicals, between fraternals, and so on.

The difference between identical twins is no greater than between repeated tests of the same individual. If the reader is predisposed to stress heredity, he will notice that the results obtained conform very well with the known differences in heredity. An environmentalist finds comfort in the very same table of results, for he points out that fraternals are certainly more alike in environment than other siblings, and possible less alike in environment than identicals. We could never reach a unanimous verdict on the strength of the evidence so far presented.

The differing environment of siblings

Environment covers a multitude of factors capable of affecting the individual's development. Food supply, poisons, infectious diseases may act either before or after birth. School and community are important, and the home deserves special emphasis. That the home is not the same for two brothers is clear from the fact that one has an older, the other a younger brother, a difference which may be quite important. One child may be mother's pet and the other father's; and this difference in companionship may amount to something in intellectual stimulation. Besides, it can fairly be said that each individual creates his own environment, in large measure, and that if two siblings differ by heredity they are sure to differ correspondingly in environment. They pick out different toys, companions, reading matter, even radio programs, and thus to be treated differently, for it is practically impossible to treat two individuals alike if they are unlike in intelligence or personality.

Evidence from foster children

Foster children deserve attention here because they are usually adopted by superior foster parents, while their own parents are not above average in intelligence or in social status. Foster children seem to be treated in general as well as own children in superior homes, and they turn out encouragingly well. But the scientific question is whether they turn out equally well with own children from superior homes. The best foster children to study are those adopted soon after birth, whose whole childhood is passed in the superior environment of their foster homes. Children who come into the hands of child-placing agencies soon after birth are commonly the children of unmarried mothers, who feel compelled to leg their babies go at the earliest possible moment. Unfortunately for our purpose, little can be learned regarding the intelligence of the own fathers of most of these children. Form their parentage we should expect these foster children to show an average IQ of not far from 100; while own children of their foster parents, or of parents of the same class, have an average IQ of 115. Two large samples of such children, taken into superior foster homes within the first year after birth, have been tested, one sample in Chicago and the other in California.

The IQ of the Chicago sample in later childhood averaged 104, and that of the California sample averaged 107. The difference between 100 and 104—107 would accordingly be ascribed to difference in environment, and the difference between 115 and 104—107 to difference in heredity. The weak point in the argument in our uncertainty regarding the heredity of these children: if it was lower than estimated, environment had a greater effect than appears from the figures; but if it was little higher, there remains no difference to attribute to environment. The only sure difference is that between foster children and own children from equally superior homes, and this difference must apparently be due to heredity.

There are several other ways of utilizing the data from foster children, and some of these ways strongly indicate a real difference due to environment. But none of the indications are entirely conclusive as yet. More definite results can be expected from the continued study of foster children as well as of twins. From all the evidence it is fairly certain that both heredity and environment are real factors in causing individual differences in intelligence. Brain disease or injury—an environmental factor—can certainly lower the individual's intelligence to near zero. Social environmental factors in the home, school and community can raise the individual's level by a moderate amount at least. There is no evidence to indicate that any environmental influences, whether physical or social, can bring the person of low heredity up to average intelligence or personality, or the person of average heredity up to a really superior level.

The differing heredity of siblings

Brothers and sisters—"siblings," as they are called

without distinction of sex—are indeed more alike in heredity and in home environment than children taken at random from the community; but they do not have the same heredity nor the same environment. To justify this statement with respect to heredity, we need one additional biological fact. Every human cell possesses 24 pairs of chromosomes, each pair including one from the individual's father and one from the mother. But the ovum in ripening prior to fertilization extrudes one chromosome from each pair—either one, just at random—and the ripe sperma-tozoon, similarly, contains only one of each pair. In the fertilized ovum the pairs are restored, but the whole combination is new and is practically certain to be different in any two fertilized ova even from the same mother. Therefore siblings do not have identical heredity.

Identical twins reared apart

If we could arrange to have one of a pair of identical twins reared in Russia and the other in France, we could safely predict a difference in language and probably in social and economic ideas. If one were brought up in the city and the other in the country they would certainly differ in their manners and in the kind of knowledge and skill acquired. But could we produce a difference in intelligence in any such way? Of late years attempts have been made to discover pairs of identical twins who have been separated soon after birth and reared indifferent families and communities. Several pairs have been found; more are needed. Here is the history of one pair. Identical twins girls were adopted by different relatives at the age of five months and brought up one in the city and the other on a farm.

The city girl had the more schooling and became a

clerical worker and also a piano teacher. The farm girl became competent in running the farm. Neither married up to the age of 28 when they were tested. The farm girl was found to be an inch taller than the city girl, 28 pounds heavier, more muscular and in better physical condition. She appeared to be more of a leader, more conversational, and fully as competent in general. In the intelligence tests, however, the city girl made the better scores. In educational tests the difference was about the same, while in a non-verbal performance test it was smaller, with the city girl still leading. The rather large difference of about 19 points in IQ must be due to environment, and the question is whether this is a real difference in intelligence.

Were the tests equally fair to city and country?—a question raised before. If we accept the usually obtained differences between city and country children as proof of genuine intellectual differences, then we can say that the country girl in this pair of identical twins was less intellectual than her city twins because of environmental factors. But if we doubt that the tests are fair to city and country, or to different races, social classes, etc., then we must discount the difference found in this pair of identical twins. This particular pair showed a greater difference than has usually been found in identical twins reader apart. Of ten cases so far reported, six show only a trifling difference between the twins, four show rather substantial differences of 12—19 points, and the average for all is 8 points. In several cases it is difficult to judge which twin had better environmental advantages.

If we should expect to find still larger differences in test results, but the question would remain whether we were making a fair comparison. This subject will continue questionable until the old slogans are laid aside and new problems formulated, such as these: What kinds of difference in intellectual performances are produced by differences in environment? What environmental factors are effective in producing intellectual differences? If identical twins reared apart can afford a measure of the environmental factor in intellectual differences, where shall we look for comparable evidence on the hereditary factor? We need to locate individuals of unequal heredity growing up in the same or equivalent environments.

8

POPULATION POLICY

Population policies are deliberately constructed or modified institutional arrangements and/or specific programs through which governments influence, directly or indirectly, demographic change. For any given country, the aim of population policy may be narrowly construed as bringing about quantitative changes in the membership of the territorially circumscribed population under the government's jurisdiction. Governments' concern with population matters can also extend beyond the borders of their own jurisdictions. Thus, international aspects of population policy have become increasingly salient in the contemporary world. Additions to the population are primarily the result of individual decisions concerning childbearing. Within the constraints of their social milieu, these decisions reflect an implicit calculus by parents about the private costs and benefits of children. But neither costs nor benefits of fertility are likely to be fully internal to the family: they can also impose burdens and advantages on others in the society. Such externalities, negative and positive, represent a legitimate concern for all those affected. The present study briefly discusses how individual and collective interests were reconciled in traditional societies, summarizes the population policy approaches adopted by the classic liberal state, and sketches government responses to the low-fertility demographic regime that emerged in the West between the two World Wars. In greater detail it considers international population policies after World War II and contemporary population policy responses to below-replacement fertility.

Population policy may be defined as deliberately constructed or modified institutional arrangements and/or specific programs through which governments influence, directly or indirectly, demographic change. The generality of the definition lends itself to varying interpretations. For any given country, the aim of population policy may be narrowly construed as bringing about quantitative changes in the membership of the territorially circumscribed population under the government's jurisdiction. Additions to membership are effected only through births and immigration, losses are caused by emigration and by deaths. Concern with this last component is usually seen as a matter for health policy, leaving fertility and migration as the key objects of governmental interest in population policy.

More broadly, policy intent may also aim at modification of qualitative aspects of these phenomena—fertility and international migration—including the composition of the population by various demographic characteristics and the population's spatial distribution. Furthermore, governments' concern with population matters can also extend beyond the borders of their own jurisdictions. International aspects of population policy have become increasingly salient in the contemporary world.

POPULATION CONTROL IN TRADITIONAL SOCIETIES

Rulers of any political unit have a stake in the size and composition of the population over which they have authority, hence an incentive to try to influence demographic change in a desired direction. Thus "population policy" may be said to have a long history, starting at least with the empires of the ancient world. Greater numbers tended to connote greater wealth and power, at least for those at the apex of the social pyramid. Measures encouraging marriage and sometimes immigration testify to the prevailing populationist sentiment among rulers throughout history. But the leverage of the weak pre-modern state over fertility in traditional societies was necessarily limited. The dominant influence setting the patterns of reproduction was located, instead, in a deeper layer of social interaction. Births, the key element affecting population change, are produced by individual couples—seemingly an intensely private affair yet one in which the immediate kin group and the surrounding local society in which that group is embedded have a material stake. All societies, if at varying degrees, grant a measure of self-sovereignty to their members. An individual has certain rights over his or her direction in life. But this is always subject to some constraints, not only biological but also social. Well before rights and obligations are formally codified in legal terms, they are established through spontaneous social interaction—a self-organizing process. Restrictions on freedom to act take the form of social expectations and pressures that individuals can ignore only at considerable personal costs to themselves.

Typically, there is a strong expectation that men and women should marry and have children. Parental and kin obligations in the matter of bringing up children are well understood by all adults and informally enforced by the community. In most societies there is the expectation that children are to be born to married couples only; that a man can have one wife at a time;

that a husband is obligated to support his wife and a father his children; and that he can expect reciprocal services from them. And informal rules shaped by community interest tend effectively to regulate the entry of foreigners. The fabric of such demographically relevant behavioural stances, supported by internalized personal norms and buttressed by religious injunctions, is a product of social evolution; how effective such institutions are becomes an important determinant of societal success. As a classic statement of the British demographer Alexander Carr-Saunders (1922: 223) put it, persons and groups of persons are naturally selected on account of the customs they practise, just as they are selected on account of their mental and physical characters. Those groups practising the most advantageous customs will have an advantage in the constant struggle between adjacent groups over those that practise less advantageous customs.

Few customs can be more advantageous than those which limit the number of a group to the desirable number....[In the traditional society] there would grow up an idea that it was the right thing to bring up a certain limited number of children, and the limitation of the family would be enforced by convention. Given the harsh biological and economic constraints pre-modern societies invariably experienced, that "desirable number" presupposed fairly high fertility: high enough to provide a sufficient margin of safety over mortality. Successful societies—societies that survived to the dawn of the modern era—thus obeyed the biblical injunction to be fruitful and multiply, even though such multiplication as a matter of historical record was necessarily very slow. But traditional demographic regimes resulting from spontaneous social interaction achieved modest growth rates at varying levels of fertility and mortality. Early modern Western Europe succeeded in maintaining a relatively low average level of mortality by means of keeping birth rates low, primarily by means of a fairly high average age of marriage and substantial proportions that remained permanently single. A contrasting pattern, such as in India, combined early and universal marriage and a consequent high level of fertility with slow population growth by virtue of death rates that were also high, approximating the level of the birth rate. With respect to the rate of population growth, these different combinations of birth and death rates in traditional societies were very similar. The potential for rapid population growth that might be triggered by a fall of mortality was, however, much higher when the pre-modern equilibrium was the result of a combination of high mortality and high fertility.

RATIONALE FOR POPULATION POLICY

Modernity—the rise of democratic state formations reflecting the public interest and the emergence of rapid economic development—brought about the realistic promise of realizing age-old human aspirations for a better life. The state increasingly came to be seen as an institution created by the voluntary association of free individuals to further their interests. The central function of the state was to produce public goods—goods that individuals cannot secure for themselves. The US Constitution, promulgated in 1789, articulated key items in the collective interest concisely and with universal validity. The aim of the Union formed by the People was, in the words of the Constitution's Preamble, to "establish Justice, insure domestic Tranquillity, provide for the common defence, promote the general Welfare, and secure the Blessings of Liberty to ourselves and our Posterity." In pursuing such goals, regulation of immigration into a state's territory is clearly defined as a public good, thus delineating a particular role for population policy. And aggregate fertility may also be construed as a public good, if its level as determined by spontaneous social interaction is too high or too low in terms of the collective interest.

The potential role of the state in regulating immigration is straightforward: individuals wishing to restrict or promote it cannot set up their own border patrols or issue entry visas. Individual preferences in the matter, however, are likely to differ. It is the task of the government to weigh and reconcile conflicting individual desires and come up with a policy deemed the best under the accepted rules of the political process. To claim a role for the state in the matter of fertility is more problematic. Additions to the population are the result of a multitude of individual decisions concerning childbearing. Within the constraints of their social milieu, these decisions reflect an implicit calculus by parents about the private costs and benefits of children to them. But neither costs nor benefits of fertility are likely to be fully internal to the family: they can also impose burdens and advantages on others in the society. Such externalities, negative and positive, represent a legitimate concern for all those affected. An individual's influence on the fertility of other families, however, is very limited: there are no private markets offering preferred patterns of aggregate demographic processes to individual buyers. Remedying such market failure may then be attempted through intervention by the state so as to affect individual behaviour in order to best serve the common good—the good of all individuals.

The earliest clear formulation of the population

problem as a problem of coordination among individual preferences, hence establishment of the rationale for potential state intervention in the matter of fertility, was given by William Foster Lloyd, an Oxford mathematician and economist, in an essay published in 1833. In the spirit of the Malthusian concerns of his time, Lloyd (1833/1968: 22–23) envisaged the possibility of overpopulation even under conditions when all families have only the children they actually want and suggested the direction in which remedy ought to be sought:

The simple fact of a country being overly populous...is not, of itself, sufficient evidence that the fault lies in the people themselves, or a proof of the absence of a prudential disposition. The fault may rest, not with them as individuals, but with the constitution of society, of which they form part. Population policy should therefore strive toward institutions and incentive systems—a constitution of society—that provide signals to individuals guiding them to behave in harmony with the collective interest.

POPULATION POLICY IN THE LIBERAL STATE

Technological progress and consequent improvements in the standard of living in modernizing societies result in a far more effective control of mortality than was possible in the traditional society. But the fall of the death rate accelerates the rate of population growth, which, in turn, could strain the capacity of the economic system to accommodate the increased population numbers. Falling living standards then would once again increase death rates, reestablishing an approximate balance between births and deaths at a low standard of living. This was the pessimistic central vision of T. R. Malthus's 1798 Essay. But this outcome, although held to be highly probable, was, according to Malthus, avoidable. Given sound public policies, there was an alternative to subsistence- level equilibrium, both agreeable and achievable.

A salient element in the 1798 Essay, and in subsequent writings influenced by it, was disapproval of the schemes for poor relief prevailing in Britain and elsewhere in Europe—on the grounds that they were likely to encourage irresponsible reproduction. Efforts of the paternalistic state to reduce poverty were held to be misguided; by stimulating fertility, hence population growth, such efforts would generate only more misery. Malthusians argued that the state's correct stance in demographic matters, as in the economy at large, was laissez faire. This would foster the prudential habits among the general population similar to those that already existed among the propertied classes. It would do so by assuring that the costs of childbearing were not shared by society at large but were primarily borne by the individual couples having children.

Heeding such a prescription did not imply that the state was to play a passive role in demographic matters. Malthus's own writings, most clearly his 1820 tract Principles of Political Economy (1989: 250–251), spell out a broad agenda which expresses the philosophy that came to be dominant in the liberal states of the West in the nineteenth century. Material improvements, such as higher wages for labour, could indeed be defeated if they would be "chiefly spent in the maintenance of large and frequent families." But Malthus also envisaged a different, happier possible outcome: "a decided improvement in the modes of subsistence, and the conveniences and comforts enjoyed, without a proportionate acceleration of the rate of [population] increase." The possibility of such diametrically different responses to the stimulus of higher wages suggests a large element of indeterminacy in fertility behaviour. To Malthus, the causes of these divergent responses were to be found in the circumstances, social and political, in which people lived—in particular, whether those circumstances hindered or rewarded planning for the future. From his analysis he derived a prescription for a population policy that would yield the hoped-for demographic outcome:

Of all the causes which tend to generate prudential habits among the lower classes of society, the most essential is unquestionably civil liberty. No people can be much accustomed to form plans for the future, who do not feel assured that their industrious exertions, while fair and honourable, will be allowed to have free scope; and that the property which they either posses, or may acquire, will be secured to them by a known code of just laws impartially administered. But it has been found by experience, that civil liberty cannot be secured without political liberty. Consequently, political liberty becomes almost equally essential. During the long nineteenth century—which may be thought of as stretching to the outbreak of the First World War—the politics in Europe and in its overseas offshoots favoured, even if imperfectly, the development of institutional and legal frameworks in harmony with such principles. This, in interaction with economic and cultural changes shaped by the industrial revolution, created a milieu that fostered the prudential habits of parents, rendering the micro-level calculus of the costs and benefits of children increasingly salient. Rising demand for labour, including greater use of child labour, and rising income levels tended to sustain high fertility or even to stimulate it.

But rising material expectations, broadening opportunities for social mobility, and the patterns and circumstances of urban living pulled in the opposite direction. This was powerfully reinforced by some programmatic activities that were consistent with the limited role the liberal state claimed in managing the economy. These included public health programs and projects aimed at improving basic infrastructure for transport and communication. And most importantly, the state, or local government, assumed a key role in fostering, organizing, and financing public education. At basic levels school attendance was made mandatory and enforced and, in parallel, labour laws curtailed the employment of children.

Reflecting long-standing cultural values and religious injunctions, and contrary to laissez-faire principles, the liberal state generally banned the spreading of contraceptive information and the sale of contraceptive devices and made abortion illegal. Such restrictions typically remained in effect well into the twentieth century. But by all evidence, any upward pressure on fertility from these restrictions was swamped by the downward pressure on parental demand for children resulting from the state policies and programs just mentioned. By the last quarter of the nineteenth century birth rates were falling rapidly in the countries of the West. In many cases, rates of population growth fell also, despite continuing improvements in mortality. In Europe this trend was facilitated by emigration, which both sending and receiving countries—notably the United States, Canada, and Australia—either positively encouraged or at least permitted. The stance of the liberal state on population policy thus brought about the prospect of a new demographic equilibrium in the West that could be consistent with continuing material progress: achievement of a stationary population at low levels of fertility and mortality and allowing freedom of movement internationally.

POPULATION POLICY BETWEEN THE WORLD WARS

The massive losses of life resulting from World War I and from the influenza pandemic in its immediate aftermath, and the sharp drop in the number of births during the war years, were temporary disruptions in the steadily declining trends of fertility and mortality characterizing the prewar decades in the West. Those trends soon made it evident that there is no built-in guarantee that the sum total of individual fertility decisions will eventually settle at a point at which, in the aggregate, the rate of population growth will be exactly zero or fluctuate tightly around a zero rate. Although, owing to relatively youthful age distributions, the rate of natural increase remained positive, by the late 1920s demographers realized that fertility rates in several Western countries had fallen to such a low level that, in the longer term, natural increase would become negative. This trend became more accentuated and more general under the impact of the Great Depression. Some observers foresaw a "twilight of parenthood." Just as excessive reproduction called for corrective public policies, there were calls for corrective action achieving the opposite result: enhancing fertility so as to assure at least the simple maintenance of the population. In some countries the ban on contraceptives was tightened and the penalties on abortion were increased. These measures had little effect. Neither did, predictably, governmental exhortation appealing to families to have more children. The most promising avenue for population policy seemed to be to use the instruments available to the state for redistributing income so as to reward demographic behaviour considered socially desirable (and to discourage contrary behaviour). By the 1930s such pronatalist policies came to be fairly widely if rather tight-fistedly applied in a number of countries. Among Europe's emerging democratic welfare states, Sweden and France were pioneers in providing financial rewards and services in kind to families with children, especially to larger families. (Sweden, however, also allowed liberal access to contraception.) Similar policies were applied with equal or greater vigour in fascist Italy and Nazi Germany.

Invariably, the proponents of such policies claimed some results in terms of birth rates somewhat higher than would have been expected in their absence. But the latter quantity is a hypothetical one, which introduces a necessary caution to such claims. More pertinently, when average fertility is low, the birth rate in any given year is an unreliable measure of long-run fertility. Couples have considerable latitude to time the birth of their children earlier or later, without affecting the number of children they ultimately wish to have. Logically, pronatalist policies seek to affect that lifetime total rather than aiming at temporary increases in the birth rate.

INTERNATIONAL POPULATION POLICY AFTER WORLD WAR II

In the countries that the United Nations categorizes as less developed, population policy issues attracted little attention until the middle of the twentieth century. Fertility remained high, more or less at its pre-modern level. Population growth was slowly increasing, however, as a result of improvements in mortality. Following the end of World War II, mortality decline accelerated greatly and as a result so did population growth. In 1950, the world population was 2.5 billion. Some 1.7 billion of that total was in countries classified as less developed, with an average annual birth rate of 44 per 1000 population —twice as high as in the more developed group. Unless a decline of the birth rate got underway fairly rapidly, an unprecedentedly large expansion of human numbers was inevitable.

Although the trigger of such population growth was a welcome development—falling death rates—growth rates that would double or even treble a population within a generation seemed a major obstacle to development. And the large and widening differential between the more developed and the less developed countries in terms of population size and average income levels was seen as holding out the prospect of major dislocations and long-term instabilities within the international system. As a result, in the 1950s an intense debate started on what policies could reduce fertility in the less developed world. This policy debate was primarily Western, much of it American, just as the diagnosis of the problem itself had been. The proposals that emerged were to be applied in countries representing a large and growing share of the global population. Population policy became international.

The West of course already had relatively low fertility, and with it much historical experience on why birth rates fell. Social science analysis was virtually unanimous in interpreting this experience. The explanation centred on the role of changing structural conditions of the economy, conditions to which micro-level units of the population tended to respond, in demographic as well as in other matters. Demand for smaller families was seen as the primary force determining birth rates; the means by which couples regulated their fertility was not unimportant, but seen as a distinctly secondary factor. If the demand was strong enough, fertility would be low, even if birth control technology was primitive. A transition to low fertility presupposed changing preferences, and such preferences were responses to market signals. If policy was to have an explicit role, it would be through reinforcing those signals. In the experience of past fertility transitions, four components of the incentive structure seemed especially pertinent: (1) the direct costs parents must incur in bringing up children; (2) the opportunity costs of children to parents, that is, the earnings a couple must forgo because of children; (3) the contribution of children to family income through labour services; and (4) the contributions of children to parents' economic security in old age, in comparison to alternative sources of security.

Fertility declines when shifts in these components make family limitation advantageous to couples, overcoming cultural resistance supporting traditional behaviour. Patterns of development generate that effect when at least some, but especially when all of the following conditions are fulfilled: (1) social expectations and formal institutional arrangements place on parents the major financial responsibility for raising their own children, including much of the cost of education and health care; (2) women have access to income-earning opportunities in the labour market, including jobs not easily compatible with childbearing and child-rearing; (3) social institutions make formal education (primary and early secondary) compulsory and effectively enforce school attendance; (4) child labour is made illegal; (5) effective legal guarantees of property rights, legal enforcement of private contracts, and the development of public and private insurance and pension schemes provide attractive and comparatively secure alternatives to children as a source of old-age security.

Social and institutional conditions that make such changes potent generators of fertility decline include the following: (1) emphasis on personal economic contribution (rather than, for example, class status or political loyalty) as the primary factor determining a person's earnings, thus providing an incentive for increased investment in human capital; (2) systems of promotion that provide opportunities for upward social mobility according to merit and tolerate downward social mobility; (3) openness to outside influences that create rising expectations with respect to material standards of living; and (4) emphasis not only on the rights but also on the social and economic responsibilities of the individual. Some of the factors that prompted the fall of fertility in the West also became potent in the less developed countries as concomitants of successful economic and social progress. International conferences, for example the 1974 Bucharest conference on population, spelled out many of the essential socioeconomic changes necessary for fertility to shift from high to low levels. But assembling the instruments so identified into a coherent development strategy of institutional-structural reforms, reforms justified also by demographic objectives, remained an elusive task. In promoting development governments came to see their roles not in supporting institutions harnessing the market, but replacing the market in key developmental tasks through specific goal-oriented categorical programs. In the matter of population policy, the rapid postwar progress in the technology of birth control provided an appealing apparent short-cut for achieving fertility decline through programmatic means. Markets, it was held, could not be relied on to bring that technology to those wishing to practice birth control. Governments could, instead, organize free delivery of birth control information, and provide effective means for preventing births to all those, primarily (it was assumed) women, who wished to plan their families. Surveys indicated that there existed a substantial latent demand for such services. Satisfied customers, in turn, would serve as role models, bringing new clients to the program.

By the mid-1960s, in programmatic terms the international population policy debate on the relative importance of demand versus supply was essentially decided in favour of the latter. For the next quarter-century, population policy in the developing world became essentially synonymous with family planning programs.

Financial and administrative limitations within developing countries necessitated the heavy involvement of foreign assistance in launching and sustaining family planning programs. Although donor countries' own domestic experience in this area was practically nonexistent, such aid, justified by the seriousness of the "population problem" (a problem whose cause was defined in terms of aggregate indexes of population growth), was readily forthcoming, partly in the form of bilateral assistance and partly through international aid institutions.

The cost of birth control technology was, in itself, relatively modest. But sustaining an effective delivery service did represent significant claims on scarce human and material resources. Declared demand for birth control does not necessarily translate into effective willingness to practice it: conflicting desires may interfere. Weakness of measures of latent demand, or "unmet need," is reflected in the requirements that programs are supposed to satisfy if they are to be successful. These typically included such items as "doorstep accessibility of quality services," "broad choice of contraceptive methods," "forceful IEC [information, education, and communication] programs," "sound financing strategies," "sound management with proper logistics," "evaluation systems," "a continuous process of strategic thinking, planning and management, " and "staff leadership for program parameters".

The effectiveness of family planning programs in reducing fertility remains a matter of controversy. According to international guidelines, programs recruit their clients on a strictly voluntary basis. By accepting the service voluntarily, the individual acceptor demonstrates that she values that service. But some of the more successful programs, notably in Asia, tended to increase acceptance by often heavy-handed methods of persuasion, and , in the especially important case of China, by coercion backed by legal sanctions. Where fertility fell in less developed countries with active family planning programs, it is typically found that program-provided free services account for a large percentage of those practising contraception. This non-surprising result is then often taken as an indicator of success in reducing aggregate fertility. But what would have happened in the absence of the program is conjectural, hence routinely ignored.

Some less developed countries that lacked government programs also experienced major falls of fertility: Brazil is a conspicuous example. Similarly, if programs have seemingly only minor success in reducing fertility, this can be taken as evidence that the program is inadequately financed, organized, and managed: greater efforts would have led to better results.

Family planning programs as they were commonly conceived had a strong resemblance to health programs. But given the special priority accorded to family planning services in foreign assistance, typically they were organized as a separate "vertical" program, or kept administratively distinct within the broad health program. The justification for such treatment was that while acceptors of family planning services are recruited because the program satisfies their individual need, the program also serves a national developmental need by helping to reduce aggregate population growth, hence deserves priority. Once a family planning program is organized, its managerial and professional cadres form a natural advocacy group strongly interested in the program's sustenance. Invoking the public interest in lowered fertility, as distinct from simply serving the needs of the clients of the program, has long served as a key supporting argument in that endeavour.

Over time, this developmental prop has eroded. This was in part a result of criticisms of the intrinsic scientific merit of the argument but most of all a reflection of the extensive decline of fertility that has occurred, a decline often attributed to the success of the family planning programs themselves. On the eve of the 1994 International Conference on Population and

Development, a review, considered highly respectful of the importance of fertility decline for successful development, reached the guarded conclusion that the data on the subject "mostly support the view that rapid population growth in poor countries under conditions of high fertility is inimical to many development goals"—with stress on the qualifying words "mostly," "high fertility," "many development goals," and "rapid," and with the last-mentioned term defined as "in excess of 2 percent annually" (Çassen 1994: 13). By that time, among world regions, only Africa and West Asia had a population growth rate meeting that criterion. Accordingly, the development rationale of family planning programs was gradually dropped and was replaced by the argument that the programs satisfy important health needs and help people exercise a fundamental human right. The Cairo conference formalized this shift: even though the name of the conference for the first time included a reference to development, scant attention was paid to that concept. Family planning programs were redefined, instead, as reproductive health programs, responding to a broader range of women's health needs, such as prevention of unsafe abortions and sexually transmitted diseases, including HIV/AIDS. But beyond this, new emphasis was put on some requirements that would contribute to women's empowerment: reduction of infant and maternal mortality and improvement in girls' education and women's opportunities for employment and political participation. Although the connection was not highlighted, these are conditions that are likely to help reduce the birth rate through stimulating the demand for smaller family size. The Cairo conference, in effect, reverted to some key elements of a demand strategy for reducing birth rates. The future of family planning service programs is thus left in a somewhat tenuous status. Without invoking a collective interest in a wider practice of birth control, it is not clear what level of priority should be accorded to such programs as just one part of publicly financed health programs, or indeed relative to many other social welfare programs that also serve demonstrable human needs. Not surprisingly, there are increasing efforts in national programs to rely on the market in enhancing access to contraceptives and to provide program services on a fee-for-service basis.

POPULATION POLICY IN RESPONSE TO BELOW-REPLACEMENT FERTILITY

During the second half of the twentieth century debates about population policy, and consequent programmatic action, were centred on the issue of rapid population growth in the less developed world. Toward the end of this period, however, a quite different demographic phenomenon has begun to attract increasing attention: aggregate fertility levels that are inadequate for the long-run maintenance of the population. Analytically, the potential population policy issue raised by low fertility is identical to the problem inherent in rapid population growth: it is caused by the disjunction between the sum total of individual reproductive decisions and the collective interest in a long-run demographic equilibrium. But this time individual aspirations generate a deficit rather than an excess in population growth. The syndrome, as was noted above, is not entirely novel: it was detectable in fertility trends in the West, especially in Europe, in the 1920s and 1930s, and in some instances, notably in France, even earlier. But in the decades immediately following World War II, the baby boom seemed to make the issue of low fertility moot. Indeed, by any historical standard, population growth was rapid during the second half of the twentieth century even in the developed world. Europe's population, for example, grew during that period from 550 million to about 730 million.

The baby boom was, however, a temporary interruption of the secular downward trend in fertility. By the 1970s, the net reproduction rate was at or below unity in most countries in Europe and also in the United States. In the US fertility stabilized at or very close to that rate, but in Europe fertility continued to decline. By the beginning of the twenty-first century, the average total fertility rate in Europe was 1.4. Such a level, if maintained indefinitely, would result in a population loss of one-third from generation to generation, that is, roughly, over each period of some 30 years. In some countries, notably in Southern, Central, and Eastern Europe, period fertility rates were at low levels without historical precedent for large populations. If continued, in the absence of large compensatory immigration this would not only lead to rapid population decline but also result in very high proportions of the population at old ages. It might be expected that in the affected countries such prospects would generate not only concern, but also vigorous remedial policy action.

By and large, however, this response has not been evident. Most governments as well as the general public tend to view below-replacement fertility with an equanimity quite unlike the alarmed reaction that the same phenomenon elicited when it first emerged between the two World Wars. And explicit pronatalist policies, common in the 1930s, are conspicuous by their absence. There are a variety of reasons explaining this indifference. First, the preeminent population issue confronting policy-makers in the post–World War II period was rapid global population growth. Programs aimed at moderating fertility in the developing world received assistance or at least encouragement from the rich, low-fertility countries. Although the rationale was modified over time, such assistance and encouragement have continued, as indeed substantial further population increase in the less developed countries is still anticipated in the early decades of the twenty-first century. Even though population issues tend to be sui generis, reflecting differences in demographic behaviour country-by-country, there was, and remains, a perceived dissonance between fertility-lowering assistance to other countries and engaging in action at home serving the opposite aims. Faulty logic notwithstanding, the international terrain has not been favourable for domestic pronatalism. Second, the natural rate of increase—the difference between the number of births and the number of deaths—is still positive in many of the countries with fertility well below replacement. This is the result of age distributions—reflecting past fertility and mortality and notably the effects of the postwar baby boom—that still favour population growth. While this momentum effect is temporary, the longer-term implications for population decline and population aging are only dimly perceived by the general public and provide an excuse for inaction on the part of policy-makers.

Third, when those longer-term demographic effects are understood, a calmer attitude still prevails. There is an inclination, reinforced by increasing concern with the quality of the natural environment, to regard a degree of demographic "decompression" as a not necessarily unwelcome prospect, especially in countries with an already dense population. And it is assumed that the economic and social disadvantages that might be imposed by a declining population can be effectively dealt with through institutional adjustments and social policy measures other than measures aiming for a higher birth rate. A demographic policy often regarded as potentially helpful in this regard is encouragement of immigration. That willing immigrants are available to compensate for low birth rates is taken for granted—a realistic assumption in high-income countries.

Fourth, there is a vague expectation that the population decline, impending or already begun, will in due course trigger corrective homeostatic mechanisms, leading to a spontaneous rebound in the level of fertility. Another baby boom may not be in the offing, but fertility may rise sufficiently to once again reach or at least approximate replacement level. Governments, it is assumed, would be ill-advised to interfere with this natural process by trying to increase birth rates and then seeking to fine-tune them at the desirable steady-state level. According to this view, a laisser-faire fertility policy is justified since, apart from broad upper and lower limits, governments are not competent to determine what constitutes an optimal fertility rate, or growth rate, or population size in any given year, decade, or even longer time interval.

Finally, even if the will were there, there is a paucity of effective pronatalist policy instruments. Exhortation from governments is not promising, and in any case unlikely to be tried in a democratic polity. Restrictions imposed on access to modern contraceptive technology are not politically acceptable; they would be also certain to prove a failure. This leaves the traditional levers of social policy: dispensing material incentives and disincentives so as to increase the willingness of couples to have children. Such incentives can be engineered by the government through fiscal measures, such as differential taxation, and/or through provision of services in kind. This approach was tried in the interwar years, but, as noted above, with at best limited success. After World War II many similar measures continued to be applied; in fact, with the steadily expanding welfare state, they were often upgraded and their scope, too, was extended.

However, they were no longer considered "pronatalist" but were absorbed within the more encompassing frames of family and general welfare policy. The new label partly reflected a political-ideological preference, but in part also the fact that some distinctive features of pronatalism—such as differential rewards that favoured large families, and non-means-tested or even regressive allocation of family and child benefits—were generally no longer acceptable.

Although the redistributive policies of the contemporary welfare state are biased in favour of the elderly and the poor, government-organized transfers to parents of children, or to children directly (such as through publicly financed day-care services and free or subsidized education often beyond the secondary school level, which lessen the cost of children to parents), are substantial in all low-fertility countries. Indeed, it is typically assumed that existing family and welfare policies sustain fertility above a level that would ensue in their absence. Accordingly, making these policies more generous—socializing an even larger share of child costs—is often seen as a means toward increasing fertility, whether as an outright policy objective or, more in the prevailing spirit of the time, as an unintended but welcome by-product. Such extension, however, is difficult, given the fiscal constraints of already over-committed welfare states. And more to the point, the net effect of family friendly redistribution of incomes and provision of services is uncertain. It is notable that in the US, where such schemes are distinctly less well funded than, for example, in Western Europe, fertility is, nevertheless, relatively high.

In recent decades, in modern industrial economies, participation of women in the formal labour force expanded rapidly. This tendency, reflecting market forces but also encouraged by government policy (partly as an antidote for deteriorating dependency ratios as the population becomes older), is likely to continue. Among the factors explaining the low level of fertility despite general material affluence, many observers point to the double burden on women of both raising children and working outside the home. To the extent that higher birth rates are seen to be socially desirable, the derived policy prescription is to adopt measures that make motherhood and women's labour force participation more compatible. The higher fertility in countries (notably in Scandinavia) where such measures are strongly applied, compared to countries (especially those in Southern Europe) where they are largely absent, suggests that enhanced compatibility (through day-care services, flexible work-hours, liberal sick-leave allowances, and the like) is an effective pronatalist policy even if motivated by other considerations. But it is far from clear whether the fertility differential so generated is high enough to bring the total fertility rate back to replacement level. Steady labour force participation of women during the childbearing years can certainly be made compatible with having one child or even two. It is likely to be far less compatible with sustaining, or even increasing, the proportion of women who have more than two children. Many career-oriented women voluntarily remain childless; many others prefer a single child. It follows that, to achieve average replacement-level fertility, the proportions of such women need to be counterbalanced by high enough proportions of women who have chosen third-, fourth-, or even higher-order births. There is little indication at present that policies directed at enhanced compatibility achieve that result.

When fertility is high, as it still is in most developing countries, it is a safe prediction that with economic development it will eventually decline, at least to replacement level. Since population growth cannot be sustained indefinitely, in the longer term, apart from out-migration, the only alternative is higher mortality. The record of the high-income countries indicates, however, that replacement-level fertility is not a necessary resting point. Once fertility is lower than that, predictions become highly hazardous.

European and also East Asian experience suggests that fertility has a tendency to settle below an average of two children per woman, hence a tendency toward sustained population decline. The question, to which no good answers exist at the dawn of the twenty-first century, is "how far below?" "Scandinavian"-style family policies may stabilize fertility only modestly below replacement—such as around a total fertility rate of 1.8. That would imply a fairly moderate relative shortfall of births compared to deaths, and population stability in rich countries with such vital rates could be fully or nearly compensated with a modest level of controlled immigration. Population aging would be then kept within relatively narrow limits, which post-industrial economies could readily adjust to. The demographic weight of such countries within the global total in the foreseeable future would continue to shrink, raising possible problems of a shifting geopolitical balance. Still, such demographic configurations would be likely to push the day of demographic reckoning beyond the policy horizons that governments feel an obligation to be actively concerned with. Pronatalist interventions would find at best a marginal place on governments' policy agendas.

On the other hand, fertility levels in the lowest-fertility countries—countries with a total fertility rate of 1.3 or below around the turn of the century—might stabilize at that level, or even shrink further, reflecting the decentralized and uncoordinated decisions of individuals and individual couples. Such an outcome might also foreshadow future reproductive behaviour in countries in which fertility is still fairly close to replacement level. This would create a qualitatively different demographic situation for which there are few precedents in modern history. It would represent a clear threat to the continuing viability of the countries affected. Compensatory immigration flows would have to be so large as to be inconsistent with any reasonable degree of cultural and ethnic continuity. Alternatively, population aging in the absence of immigration would create virtually unsolvable challenges, and there would be a likely drastic loss of relative geopolitical status. Spontaneous homeostatic mechanisms may not come into play to save the day, or may do so too sluggishly to matter. A radical rethinking of fertility policy would then become a necessity for social—and national—survival.

HEALTH AND POPULATION

POLICY INITIATIVES

Over the last fifteen years India has witnessed a sharp decline in the state's commitment to public health. Thus today our country has the fifth lowest public health expenditure in the world. As the National Health Policy admitted, this is, at 0.9 per cent of the GDP, lower than the average in even Sub Saharan Africa. Along with decreasing state spending on health, increasingly policy measures have encouraged the growth of the private sector in health care so that today we have the largest and least regulated private health care industry in the world. Evidence from across the country indicates that access to health care has declined sharply over this period. The policy of levying of user fees has impacted negatively upon access to public health facilities, especially for poor and marginalised communities and to women. As health care costs have increased sharply, it is not surprising that medical expenditure is emerging as one of the leading causes of indebtedness. At the same time, this has been accompanied by policies that have reduced access of the poor to public distribution systems of food so that per capita availability of food has shown an alarming decrease.

It is thus not surprising that in addition to starvation deaths, the huge load of preventable and communicable diseases remains substantially unchanged. Infant and child mortality take an unconscionable toll of the lives of 22 lakh children every year. We are yet to achieve the National Health Policy 1983 target to reduce the Infant Mortality Rate to less than 60 per 1000 live births. More serious is the fact that the rate of decline in the Infant Mortality Rate, which was significant in the 1970s and 80s, has remarkably decelerated in the 1990s. 130,000 mothers die during childbirth every year. The NHP 1983 target for 2000 was to reduce Maternal Mortality Rate to less than 200 per 100,000 live births. However, 407 mothers die due to pregnancy related causes, for every 100,000 live births even today. As per the National Family Health Surveys in the last decade, the MMR has increased from 424 to 540 maternal deaths per 100,000 live births. Partially as a result of population policies, the disincentives and the two-child norm contained in them – at variance with the National Population Policy (2000) and the commitments made at the ICPD in Cairo – there is a massive shortfall of girls in the 0-6 years age group due to Sex Selective Abortions (SSA). Violence against women has grown, and taken many new forms, including a huge increase in so-called "honour killings". Indeed it would be no exaggeration to state that population policies have added to this violence.

HOPE FROM UPA GOVT

It is thus with hope that we looked forward to the United Progressive Alliance government to initiate policy measures to arrest these trends. Although the Common Minimum Programme was committed to a substantial increase in health spending, this is not evident in the financial allocations made: the budget outlay for 2004-05, adjusted for inflation, shows no increase in the outlay for health and an 11.9 per cent increase for family planning, clearly indicating skewed priorities.

However, more alarmingly, the single line, "A sharply targeted population control programme will be launched in the 150-odd high-fertility districts", from the CMP has been acted upon to unveil policy measures for 209 districts in the country. This is a truly unfortunate move, with grave consequences for thousands of women and children from the poor, marginalised communities, especially dalits and adivasis, who along with deprivation, suffer higher levels of morbidity and mortality, and a high unmet need for health and family planning services.

REGRESSIVE PROGRAMME

These 209 districts in which the "sharply targeted population control programme" are to be launched are precisely the same districts with poor indicators for social development, especially female literacy, infant and child survival, maternal morbidity and mortality and other indicators of human and gender development. Instead of a package of health and development measures, what is being proposed is concentrating on a sharply targeted population control programme in these districts in five states. This profoundly regressive policy relies on targets for sterilisation, coercive incentives and disincentives, and massive subsidies to the private sector, recalling the worst days of India's family planning history. Not only do such measures violate basic human rights, they have also been shown to be demographically unnecessary in bringing down population growth rates.

The declining Child Sex Ratio (CSR) is one deeply worrying indicator of the outcomes of such short-sighted population policies. Between 1991 and 2001, in urban areas, the CSR has declined from 935 to 903 and in rural areas from 948 to 934. More ominously, between January and June this year, in Delhi the Sex Ratio at Birth indicates 819 females being born for every 1000 males; in the prosperous and educated South Delhi zone, where demographic transition has by and large been completed, only 762 females were born for every 1000 males. A recent study by the Ministry of Health also indicated the dolorous outcome of the imposition of the two-child norm for contesting elections. A large majority of those disqualified on this ground were dalits, the adivasis and women from poor families, defeating the very purpose of democratic decentralisation. Further, the study indicated that this norm had acted as an incentive for SSA. Clearly then population stabilisation in this form cannot be the goal since it leads to profoundly unbalanced populations.

We therefore demand that the programme measures for these 209 districts be unreservedly scrapped. We cannot have a population policy that does not hinge on equity and gender justice. There should thus be no National Population Mission. Issues concerning women's health and reproductive rights can only be part of a larger package of a health and social development policy.

We welcome the UPA's commitment to increasing state spending on health, but this should be entirely devoted to strengthening the universal and comprehensive primary health care (PHC) system. Increasing state spending cannot become the vehicle to increasing public subsidy to the private or NGO sectors, which in fact require regulation. It is evident from the CSRs that the private sector has played a deeply regressive role in health care provision.

The Tenth Plan proposals for health and nutrition need to be reviewed keeping epidemiological priorities in mind. A basket of technologically determined vertical programmes cannot substitute for a systematic strengthening of a comprehensive, universal, integrated PHC system.

Ten years after Cairo, if the commitments made there

have to have meaning, it is clear that we cannot have RCH without PHC; nor indeed can we have gender-just population policies without the enabling conditions of health and development. What we have demanded is the minimum and non-negotiable.

9

CONCEPTS OF GROWTH AND DEVELOPMENT

Life of human being never remains same. With the time, various kinds of changes take place in various aspects of the life. Various kinds of changes can be seen in the body of human beings. Not only external, various kinds of internal changes take place in a human being's body. With these changes, a lot of physical as well as psychological changes take place in the human beings. These kind of changes are described by the terms of Growth and Development.

Concepts of Growth and Development

Growth

The process through which body increases in size and weight is known as growth. It is with the help of this process that an individual's body becomes larger and heavier. Actually, when we speak of growth of human body, it results from the growth of muscles, brain and of the skeleton. Various kinds of physical changes that take place in human being as a result of conception to maturity are parts of the concept of growth. An increase in mass of the body is termed as growth. Thus, a quantitative increase in the size and shape of the body is termed as growth. When these organs changes in size and shape, it is known as physical growth. With this process, all the organs of the body got enlarged in size, they become longer, thicker and heavier in weight.

Meredith has defined the concept of growth as the entire series of anatomical and physiological changes that take place between the beginning of the parental life and ends after death of the individual.

Development

Concept of development is different from that of growth. Under this process, various kinds of changes take place in the human being which help him to become mature. These changes take place as a result of maturation and experience. Generally, various kinds of internal changes that place in an individual are included in the process of development. These changes can be qualitative as well as quantitative in nature. Development consists of various kinds of structural and functional changes which can be of positive as well as negative nature. These kind of change take place in any phase of the life. Development take place in an individual in the early phase of life, the stability which is generally found in young age and the deterioration in the capacity of individuals during old age are important constituents of the concept of development. Development is a phenomena with a wider scope, which consists of different kinds of changes taking place in an individual.

Different Stages of Growth and Development

Science has developed to very large extent and it is because of the studies done by scientists that the manner or process in which human beings take birth is being presented by the scientists. According to their studies, when two germs from the body of partners indulging in relationship gets united, a single cell is being produced, which is being called the zygote. In a very rapid pace, this cell gets multiplied. It took around 7 to 8 months to develop properly, and it is after this long period of time that the weight of this cell can be measured.

A child grow in a very set or sequential pattern. One can easily divide this sequential pattern into some parts, which are as follows :—

Pre-Natal Stage,

Post-Natal Stage,

Infancy,

Childhood,

Adolescence,

Adulthood, and

Old Age.

Various kinds of changes take place in an individual during these stages, which have been discussed in the following manner :—

Pre-Natal Stage : It is during this period or stage that two cells get united from the bodies of people of two sexes and give birth to a new cell, which is called Zygote. In the beginning, this cell has a very small size and it begins to accelerate in a very slow speed. It takes around a period of two weeks to get attached to the uterine wall and this process is being called the embryo. It is in that stage that mass cell gets divided into three layers, which are called the Ectoderm, Mesoderm, and Endoderm. All these layers grow in different organs of the body. For instance, ectoderm grows into the skin, nervous system, eyes, tooth and oral passages. This layer is the outer layer of the cell. After this situates the Mesoderm, which is the middle layer of the cell. Skeletal, kidneys, muscular tissues and blood are the main organs into which this layer grows. The innermost layer is called the Endoderm, which generally grows in digestive system and some special cells of thyroid and parathyroid glands. At this stage, cell grows at a very rapid speed and after a period of six weeks, fetus stage comes into being. During this stage, rate of growth of the cell increases to a considerable extent, and it is during this period that cell gains weight. After a period of three months, child becomes able to move in some manner. By seventh month, various organs of the body of the child gets developed. It is only in the ninth month that blood passage of blood cells gets directed into fetal blood streams. Placenta begins to break down in the ninth month and mother's blood flow in the child's blood stream at this period. Various kinds of genetic as well as environmental factors affect the pre-natal stage.

Post-Natal Stage : Parturate stage begins after the birth of the child till the severance of umbilial cord. It is after passing from this stage that child gets detached from the body of the mother. Upto this period, muscle fibres of the child do not get developed properly. At the time of birth, brain has a weight which is being equitable to the one-eighth of the entire weight of the body. After birth, period of around two to four weeks is called Neonate stage. It is during this period that various organs of the child's body adapt themselves according to the outer environment.

Infancy : The period until child attains the age of two years is known as infancy. It is during this period that a child acquires various kinds of capacities at a rapid speed. Various organs of the body gets developed at a constant speed during this period. All the children do not grow at the same speed, but pattern in which they grow is quite the same. Various kinds of hereditary and environmental factors affect the growth of the body of the children. During this period, bones of the child develop at the considerable pace.

At this stage, fatty acids present in the body grow up as a result of which child's body weight gets increased. With the time, bones and muscles of the body gets developed because of which the process of increase in body weight also gets increased. Height of child's body at this stage varies from 9 to 20 inches. In sixth or eight month, first temporary tooth starts appearing out of the gums of the baby. In comparison to upper teeth, lower teeth appear first.

During this period, nervous system of the body gets developed which result from the development of immature nervous cells. These nervous cells are being present in the human body at the time of birth. In the initial stage of development of brain, cerebrum and cerebellum gets developed. During the first year, weight of cerebellum gets triple, which plays an important role in bringing a kind of balance and postural control of the body. Body gets grow at the rapid pace in this period. Child gets to learn various kinds of activities, like walking and talking. To the outer environment, his circulatory and digestive systems get acclimatized, and during this period, all these systems start to function.

With the development, functions of the brain becomes more organised. Various kinds of voluntary motor behaviours take place. He becomes able to sit, creep and crawl and then after some time, he learns to walk without the help of anybody. With the time, an improvement comes in the various motor skills, which leads to a kind of cognitive development in the body of the child. With development of cognitive capacity, a child become more aware of himself. This kind of development of his cognitive capacity takes place as a result of his interaction with various people of the society. Social world of the infant is being dominated by his parents and other members of the family. After third month, muscles of the eye body gets stronger because of which they can move in a co-ordinated manner. By this time, their hearing ability also gets improved. It is at this stage that the process of socialization gets started in the child's life. Upto this time, they begin to turn their head, smile, and wave their arms in response to activity of some other person. They recognise the fact that they have been left alone by his parents because of which he starts crying. They begin to differentiate between the emotions of love and anger. After an age of one, they begin to respond to simple commands of yes or no and by the end of second year, they starts to take part in simple activities of adults. Thus, it can be said that this period in the life of a child is a period of self-assertion.

Childhood : Period in which the relative dependency of infant is over, is called the childhood. The time span of this period is very long, as it exist upto the period until when the child attains sexual maturity. Generally, this period differ with regard to the sex of the child. In boys, this age begins from two years and last upto fourteen years, however, in case of girls, this period ends upto they attain the age of thirteen.

Various kinds of physical and psychological changes take place within an individual's body. This period is being divided into two categories, namely, the early childhood and late childhood. From second year period known as early childhood begins and this lasts upto an age of sixth year. After this age period begins the period of late childhood, which extends upto the time when the child becomes sexually mature. During the whole period of childhood, a child growth at a uniform or considerable rate. Higher centres of brain of the child gets developed at a slower rate, however, during these years, it is being developed progressively.

Adolescence : This word has been extracted from the Latin word adolescence, which implies to grow into the stage of maturity. In reality, adolescence is a process and not a period. During this period, a child achieves the attitudes and beliefs which he need for participation in the society effectively.

In simple terms, the stage between childhood and adulthood is known as adolescence. This age begins from 13 to 19 years of age. This period consists of stress and strains of various kinds. First stage of adolescence is known as puberty, during which a child initiates his steps towards adulthood for the first time. It is during this period that primary sex characteristics of a child gets developed. In comparison to boys, girls enter this stage earlier. In case of boys this age is between 11 to 15 years, while in the girls, it is between the age of 10 to 14.

During the early adolescence age, a child is grown in a very disproportionate manner because of which this period is known as the age of awkwardness. Those adolescents are pre-occupied about their physical appearance who are adjusting to their growth spurt. The age in which girls and boys attain an increase is height is also different. In girls, this age begins from the age of nine while in the boys, it gets started from eleven, however, in all the children, changes in weight follow the same kind of pattern, which is being related to the height.

Adolescent spurt of growth is found in proportion to the magnitude for weight rather than for height, which is because of the tendency which is being found to accumulate large quantities of fat in sub-cutaneous tissues. There are various factors which give rise to this fact. Because of increase in weight, skeleton gets broader and massiveness of individual bones of the human beings get increased. This kind of increase in the stockiness of bones is more apparent in boys in comparison to girls. At this time, muscles of the body grows more rapidly. Various studies have proved that during this period, between the age of twelve to sixteen, strength of a child gets doubled, however, it reaches to the maximum level after utmost growth in height and weight. Continuous moderate improvements are being shown by motor abilities, however, this kind of improvement is more rapid in the periods of pre-pubescent and post-pubescent. During this period, physical and physiological abilities of a child gets developed to a lot of extent, which result in growth of mental abilities, with the help of which an individual can deal with various abstracted matters. During this period, mind of the child gets its own metamorphosis.

Abilities to concentrate properly and to keep the memories of various events also increases during this period. A child becomes more self-aware during this period of age. Various internal organs also gets developed. Size of the heart and blood vessels grow in size during puberty. Organics of various important systems of the body becomes more developed, which improves their ability considerably. All the changes help in improving the functioning of sex glands and the pituitary glands. Changes during this period affects the social life of a child also. They become emotionally more mature, and begin to participate in decision making processes. The impact of this period on all the children is not same. Some becomes tensed and feel stress, while some of the children get the opportunities to grow and develop their physical, intellectual and social abilities.

Adulthood : Adult life is the period for which infancy, childhood, and adolescence are considered to be the preparatory phases. When a child enters in this age group, he is considered to be prepared to face various kinds of challenges of life. This stage is considered to be the age of effective cognitive functioning. It is during period that a child has to decide or choose the proper profession in which he wants to enter.

It is during this period of life that a child gets confidence and ability to work independently. Various responsibilities are being laid upon him and he has to face various kinds of commitments. Process of aging takes place on a continuous basis, and when an individual attains the age of upto forty, a marked decline in seen in his mental and physical capacities.

Old Age : When an individual attains the age of sixty, he is considered to be in the old age. Generally people in these age group are known as senior citizens. During this period, in various ways, an individual capacities get deteriorated and declined. Various kinds of problems are being faced in doing some functions, however, by indulging continuously in some important kinds of exercises, one can cope with all these kinds of problems, however, to some extent.

Process of Growth and Development during the Period of Childhood

The span of life which begins with the cessations of infancy is known as childhood. This period begins from the age of two and extends upto a child gets into the stage of adolescence. At the end of this period, a child becomes maturer in all the respects. Thus, it is a long period of time, during which various kinds of changes take place within an individual. Into various categories this period can be divided, which are as follows :-

a. Early Childhood : At the conclusion of infancy, this period begins. In this period, growth of the body takes place at a slower rate. Process of growth does not get stopped entirely in this period, but pace of it gets reduced or somehow becomes static. Various kinds of changes are being shown in body proportion and the immatureness look starts disappearing. Manner in which the skeleton develops result in changes which occur in the body proportion. Change in body proportion is inevitable and most obvious change taking place in the body is of the lengthening process. Shoulders of the individual becomes broader and his hands and feet gets increased in size. Body of the individual becomes lengthy and it becomes thinner. When a child enters into this period, all the baby teeth of him gets developed. The total number of such teeth are 20 in number.

During this period of life, emotions are intense and temper tantrums occur at a frequent period. During the age from two to five years, perception of shape progresses at a rapid rate. It is at the age of three years that child learns to make various kind of judgements.

An individual has the capacity to learn various kinds of skills during this period. During this period, young child tends to learn the different kinds of skills as he has a very pliable body and he can adapt himself according to the needs and requirements. The kind of opportunities he gets determines the activities he will learn.

A child from poor family takes lesser time in learning skills in comparison to child from wealthier or richer families. If a child is being brought up in an over-protective environment, then his abilities to learn skills decreases considerably and vice versa. During this period, process of language growth take places at a considerable pace. Vocabulary of the child increases to a lot of extent and this helps him in expressing his emotions and feelings more easily.

b. Period of Later Childhood : This period begins when the child is around five or six years of age and it comes to an end when he attains the age of thirteen or fourteen. Sometimes, this period or age is also called the school age. In this period, one gets grow at a very leisurely rate. After the age of six, pace of physical growth comes to a decline and child loses his soft contours and top heavy stature. The growth rate of bones become slow and this tendency continues with the increase in the age.

During the early phase of this period, arms and legs grow at faster rate in comparison to the trunk of the child. The height of the child increases considerably. Body proportions of the child becomes quite same as an adult's, Child's limbs become lengthen, which brings various kinds of changes in his physical structure. Baby teeth are being over-thrown by the permanent teeth and various kinds of facial configuration take place in the child. Bones become longer as well as harder. Heart, and blood pressure of the child increases, while a decrease is found in his pulse rate. Hunger of the child gets increased during this age and children of both the ages become stronger.

Throughout this period, neural conduction rates and reaction speeds up. Child's reaction time decreases with increase in the age. Child becomes better co-ordinated and integrated and becomes capable of learning and executing more complex motor activities. This kind of co-ordination refined further during this period of time and it is being used in various activities.

Child become able to acquire adequate conception of himself as he gets a realistic conception of his environment. Size of child's conceptual world also increases. He can perceive efficiently and rapidly. They become more efficient in organising perceptual information in coherent ways. It becomes possible for an older child to process complex information.

Various kinds of new mental abilities got aroused in the age of six or seven. This period is considered to be very important for teaching the child new skills and knowledge. These kind of knowledge and skills will help him in his near future. He will become efficient in drawing various kinds of logical conclusions. His capacity to remember things increases.

A considerable growth is seen in the vocabulary and syntactic understanding of the language between the age of five and ten. Child becomes more perceptive in using the rules of language while communicating with someone. Till now, the main school from where child can learn manners and other social values remains his house. He considers his parents as the moral teachers. Thus, he is guided by his parents during this age group. Child's behaviour gets influenced by the parental behaviour and attitudes. Only that child can become confident who can have, beside his social interactions, a secure environment at home. In providing direction to the behavioural patterns to the child, home environment plays a much important role. These behavioural patterns, with the time take the shape of habits, and affect the personality of child to a lot of extent. Character of the child depends to a lot of extent on the kind of environment he gets at his house. On the child's personal and social adjustment, parental acceptance or rejection has very important influence. If child do not get support or love from his parents, he becomes emotionally insecure and unhappy, while those children who are being accepted by their parents feel emotionally secured and this provides them with the necessary courage and self-confidence. Thus, personality of a child is being affected to a lot of extent by the kind of learning situations he found at the home.

As said earlier that in the adolescent age, a child becomes self-dependent from the dependency he had earlier on his parents and other elders around him. In this period, a child becomes quite responsible and self-reliant. Thus, because of all these, child has to make various kinds of psychological adjustment through which he can adjust himself in the changed conditions, which are absent earlier.

If it is said that adolescent is the agc in which a child do various kinds of experiments and adventures, it will not be entirely wrong. There is a tendency, which is being found in the adolescent that they like to behave contrary to accepted standards which are being set up by the adults. Some children becomes mature early and they set the goals they have to achieve, while some do not indulge themselves in any kind of planning, because of which they become maladjusted to the changed needs and conditions. At this stage, various kinds of problems are being faced out by the children, because of which this age is sometimes called the problem age. Only that child can successfully cope with these problems who will be given proper kind of responsibility at the proper time. Adults should provide the adolescents with adequate opportunity to make their judgements and solve the problems themselves.

Various kinds of new situations take place in the life of children during adolescent age and child who is over-protected and who is not prepared to face various problems arising during this period can become emotionally, socially and psychologically unbalanced. During this period of life, an individual tries to find himself as a person. If planned properly, one can very easily and comfortably enter into the state of adolescence.

Categorisation of Adolescent Age

This age period can be classified into three categories, namely, the pre-adolescence, early-adolescence and late-adolescence. All these periods have their own features and characteristics, which are discussed as follows:—

Period of Pre-Adolescence : During this period, childhood personality gets broken up and various kinds of changes take place in the personality of the child. All these changes lead the child towards period of adulthood.

In this age, a child has to cope with various kinds of problems which arises because of various changes taking place in his life. Generally, children in this age group are found to be restless and hyper-active. They seems to be rebellious in nature. They have to make various kinds of adjustments in order to adapt themselves according to the new or changed conditions and emotions. One's perception regarding the parents, peers and the self gets changed during this period of time. If children are not treated properly by the elders during this period, they can respond to them with irritation and distrust.

Children in this age group are emotionally very sensitive and sometimes become over-powered with their emotions. Until now, they do not learn the skills to get alongwith their adults. A kind of conflict is being found in between the children and the parents, and this makes them frustrated. He begins to argument with his own parents if they do not provide him the authority to take his own decisions. Children during this age group requires independence and they do not want others to interfere in their life. Pre-adolescent children want to enjoy full authority in all the matters. Sometimes, they feel guilt or shame because of their behaviour towards others, but it happens seldom. They have their own ideals and values and they misbehave or even ridicule the people who do not cherish their values. They like to come out of the jurisidication of house and desire to maintain friendly relationships with a large number of people. They like to participate in the activities in which children of his age group participate. They like the exposure to mass media of various kinds.

Period of Early Adolescence : During this period, various kinds of changes take place in a child, not exactly physical but in their feelings or psychologically. They provide more importance to their body, looks and stature. As known, various kinds of physical changes take place in a child during adolescent period, which leads a kind of self-rejection in them. In this age, chief criterion of social acceptability is considered to be the physical attractiveness. They find it interesting to know more about matters relating to sex. They feel anxious and curious if they do not get this kind of information. They begins to attract towards the people of other sex. They try to find out their self-identity. In achieving personal identity, becoming independent of parents and gaining acceptance of peers are important steps. Another important step in this process is to make a choice of and do various kinds of preparations for a career, which is being decided by the child.

Period of Late Adolescence : By this period of time, most of the children get entered into their career. They get the independence from their parents and other elders around them. They try to make their own perfect world in which they try to abolish all kinds of inequalities. Till the child gets entered into this phase of life, his outlook towards the people of other sex gets changed considerably. It is during this age that children become mentally mature. This period of age is also called the period of heightened social consciousness. During this period, adolescents provide special attention to the status and social acceptability. The desire to be popular in the society is dominating during this period of age. Only that adolescent can adapt himself in the changed condition who can appreciate the personality of others and can various kinds of changes in himself which can help him in maintain co-operative relationships with the others.

During this period, body image of the child gets changed, social roles which he is required to play also gets changed and sex role gets changed to a lot of extent, which create various kinds of problems in finding out the self-identity. As he gets certain independence from the parents, he tries to make his own values and beliefs. At various times, a controversy is found between the child and his parents. This results because of various transitions which take place during this period. He becomes mentally, psychologically and socially very mature which helps him in taking his own decisions and if he does not get such independence or authority from the parents, he becomes a rebel. He behaves or does things in the manner he likes and even does not consider what his parents or other elders say. Because of this, child feels inconsistent and confused within the world of adults, where he is not given the right to involve in decision makings.

Various kinds of problems are being faced by children in this age group. These problems are generally of the psychological nature. It is very necessary to solve these problems as soon as possible otherwise, they can take the shape of serious problems. Some of the problems which an adolescent face are as follows :—

a. Children in this age group wants complete independence. They do not like any kind of boundations from his family or parents. They want to take their own decisions independently. Generally, during this period of life, a lack of understanding is being found between a child and the parents. Parents do not recognise the changes which are taking place in the child and they try to impose on them their own wishes or decisions, but this irritates the child and he becomes a rebellion.

b. Various physical changes take place in a child during this age group and the problem arises because of the fact that these changes do not take place in a simultaneous manner, because of which, sometimes, children consider their physique as improper. This kind of feeling creates a lot of problems or depression among the adolescents as their find themselves deviated from the sex appropriate physique.

c. Various kinds of changes take place in the school environment of the adolescents. During these changed conditions, it becomes quite difficult for them to adjust themselves. Now each subject is being taught by different teachers and for this reason, children are required to adjust themselves according to the personality traits of all the teachers. At one time, they may be required to obey the orders of an authoritarian teacher and at the other time, they may be required to participate in the classroom planning. Today's world is a world of competition, and an adolescent should know the skill to make adjustments according to the failure and success. At this stage, various kinds of fears kept on brooding among the adolescents, such as fear of failure, fear of disapproval of teachers and fear of disapproval of peers.

d. During this period of age, a tendency takes place in a child according to which he wants to spend his life according to his desires. He does not want to obey the rules and norms set up by the society or the others on him. But, sometimes, the norms set up by the child do not seems appropriate to the new social life, which creates a lot of problem for him. In this kind of situation, he finds himself mis-adjusted in the society's values.

e. In our country, generally it is seen that a child is treated in different manner at home and in different in outside world. Sometimes he is provided with heavy responsibilities as he is very mature while sometimes he is over-protected as he is an innocent child. This kind of dual behaviour give birth to a kind of confusion in the mind the adolescent.

f. Adolescents try to find out their identity in this period of life. Some regard them efficient in facing various kinds of problems, while some consider them incomplete to do the same. They set up certain goals for their life and if they do not get them achieved, they feel distressed and stressed. A lot of psychological stress is being put on them because of the goals they have set out for the future.

Thus, various kinds of different problems are being faced out by the adolescents in this period of age. Only that child will become more able to solve the problems generally aroused during this period of time who are provided with proper care as well as freedom by the parents. Under guidance of self-disciplined adults, transition from childhood to adult status should be preceded in a gradual manner. Only this kind of development will be accepted by the society. If liberty is being provided to them suddenly, then they will become confused and can use their power in wrong manner. This kind of thing can make them socially maladjusted and emotionally unbalanced.

Physical Education and Problems of Adolescents

The children in adolescent period like to perform or take part in various kinds of physical activities. By taking part in the physical activities, they get various kinds of personal advantages, which help them in developing their personality to a considerable extent. An adolescent can engage himself in various physical activities through the field of physical education and sports.

To become a whole or complete man, one should have the quality of self-confidence and self-control. By becoming master in some physical skill, one can get self-confidence. All kinds of sports and games help in attaining some level of self-confidence in a child, however, different activities provide this quality in varying degrees. Activities offering a definite challenge to courage contribute in making self-knowledge to a considerable extent.

Man is considered to be a social animal, and it is necessary for acquiring emotional balance to fulfil his social needs or requirements. Two important social virtues are loyalty and un-selfishness. These two qualities can be developed in a child through the participation in team sports. This kind of games or sports should be given prominent place in the curriculum of the adolescents.

If any child is lacking in some specific quality, he can fulfil it by joining certain sports activity. For instance, if one lacks in keeping close co-operation with others, he should be included in some team sports, i.e., cricket, where will be required to play while maintaining a co-operation with other players of the teams.

By participating in the games and spots, a child can develop his total personality. He becomes physiologically as well as psychologically healthy, which helps him in solving various kinds of problems which can be of different nature.

If during this period, any child gets habitual to bad habits, he will remain to practice those habits throughout the life. These bad habits are required to be detected at proper time and it is also very necessary to correct them and an important means to correct these errors is by making the child participant of certain sports activity. Through various sports and games, bad habits of posture or function can be corrected.

Various kinds of opportunities are being provided to the children for acquisition of aesthetic satisfactions by participation in the physical activities. There are some sports events which are being organised in the natural surroundings which help the children in getting closer to the nature. Such kind of aesthetic contributions play an important role in the development of an adolescent's personality.

A large number of children are found who lack physical co-ordination. This kind of lack of co-ordination can be improved by the participation in physical activities and sports. Progressive course of training can be used on such kind of children, which help in correcting the fundamental weakness which is being found in them. This will influence the performance of the child on the other activities positively.

From the above discussion it is clear that physical education and sports play very important role in solving the problems which are generally faced by adolescents. Physical activities help them in becoming more maturer and also to cope with the difficulties generally faced by them properly.

Concept of Personality

Personality is a complex phenomena, which means

different to different persons. For a common man, it is nothing but the general physical appearance of an individual, while if seen from psychological point of view, it is the manner in which an individual sees the world and the things around him. It is this complex nature of the concept that various psychologists have done deep study relating to this concept and have defined it in various manners.

True Meaning of Personality

In the general concept of personality, it is nothing but an individual's manner of living. The way in which an individual spends his life, wears the dress and interact with other individuals is known as personality. But, this concept of personality is entirely different the psychological concept. According to the psychologists, personality is not only how an individual looks or wears dress, but it is very wider in nature. In reality, from the word persona, word personality has been derived, which is a Latin word. Word persona implies the mask, which was being worn in the ancient Greece by the actors with the purpose to hide their identity while they used to take part in the theatrical plays. To present the character, to whom actor was being portrayed that he used a mask. From their the concept of persona came into existence. Thus from the origin, it can be said that word personality implies effect one exerts on the other peoples around them.

Structure of human personality is very complex, which comprises of various emotions, habits, motives and thinking pattern of the individual. All the aspects of an individual are included in his personality. One's perception, imagination, attitude and habits are being reflected by the personality and all these are important constituents of an individual's personality.

Various biological and cultural heritages or behaviours affect an individual's personality. With some biological heritage, a child comes in this world and his personality is being moulded by the cultural environment in which he is brought up. Interaction of biological organism with the social environment affects the personality of an individual to a lot of extent. The manner in which an individual adjusts himself according to the external environment is what known as personality. Thus, the manner in which one responds to the outer environment is known as one's personality.

Every individual has an individual system which reflects his behaviour and the manner of thinking. Peculiar blend of characteristics that make a person unique are known as personality. Various features make-up one's personality. These features are external in nature that interact in different manners. Personality is the organisation of all the features found in a human being. The manner in which these features or traits are integrated in an individual is different as found in the other individual. Because of this reason it is said that personality of an individual is unique in nature.

Psychological make-up of an individual is revealed through the personality. In simple terms, it can be said that it is the quality of an individual's entire behaviour. Personality is a continuous process of learning under which an individual learns various modes of responses. All the factors which make up an individual are important constituents of his personality. It consists of one's psychological make-up, his ability to adjustment himself in the changed environment and the manner in which he behaves in different environments. It is not possible to define all the criterion or features which can help in making up a good and attractive personality, but self-acceptance, acceptance by other members of the society, and performance of work efficiently are some of the important criterion which can help one in developing an attractive and good personality.

One of the most interesting areas of psychology is personality. This topic has remained an important subject area for the researchers and various researchers have defined this concept differently. Some of the definitions provided by the psychologists are as follows:—

In the words of Dashiell "Personality is the whole picture of one's organised behaviour, which can be featured by his fellowmen in a consistent manner."

According to Munn "Personality is nothing but the integration of an individual's structures, manner in which he behaves, his interests, attitudes, abilities and capacities."

In the words of Gordon W. Allport "Within the individual, personality is the dynamic organisation of psycho-physical systems that determine the manner in which individual have adjusted himself to the environment."

In the words of Ulrich "Sum total of an individual's values and attitudes and all of his traits is known as the personality. The resultant of all these traits have a dynamic nature. In other words, personality is dynamic in nature.

According to J.B. Watson "Sum of all the activities which can be discovered by actual observations over a long period of time which can provide with reliable information is known as personality."

In the words of Carl Pearson "Personality is the effect of an individual' appearance and behaviour upon the persons coming in contact with him."

According to Buskist and Gerbing "Personality is a particular pattern of behaviour and thinking which prevail across time and situations that help in differentiating one individual from the other.

From the above definitions, one can grab out various features of one's personality, some of which are as follows:—

a. Product of Heredity and Environment : With various hereditary biological traits a child is born but by the socio-cultural environment in which he lives, his personality gets influenced. Through the interaction of heredity trait with environmental factors, personality of an individual gets affected to a lot of extent. Because of this it is said that an individual's personality is the product of hereditary and environmental factors.

b. Adaptable to Changing Conditions : Personality of an individual helps him in adapting himself to the environment or the conditions in which he is required to lead his life. With change in conditions, personality of an individual also gets changed. With this various kinds of changes or modifications take place in the behaviour and habits of the individual.

c. Components of Personality are Permanent in Nature : All the components which make up an individual's personality are more or less static or permanent in nature.

d. Everyone has Unique Personality : Personality

characteristics found in an individual are different as being found in other individual. These characteristics help in differentiating one individual from the other. No two individuals can have similar personalities and personality can be specific or unique to an individual.

e. Learning affects Personality : An individual's personality gets affected to a lot of extent by the his learning and the experience. The impact of learning on an individual can be evaluated from his personality.

f. Development of Personality is a Continuous Phenomena : Throughout the life of an individual, his personality keeps on developing. There can be no stage in an human being at which his personality can be said to have developed fully. It is because of this reason that it is said that personality of an individual has a dynamic nature and not a static one.

g. Various Constituents of Personality are Inter-Related to each other : Personality of an individual is made up by various components, but these components do not function in isolation, but they are inter-related and dependent on each other. It is the unified form of all the components that constitute the personality of an individual.

10

FOOD AND NUTRITION

Introduction

One of an important feature of found in all the living beings is that they experience hunger for satisfaction of which they have to take consuming some material by which they can get their stomach filled. Here it is important to note that mention has been made of all the living organisms as it is not only the human beings who sense such feeling of hunger, but all the animals and insects existing on the earth in any form. As human beings are considered to be superior creature of the god, thus, for satisfaction of their appetite, they first get such food items ready for consumption, while other animals and insects consume food items in their raw form. Today, various techniques have been devised by which human beings add taste to the food items. Not only this, for this purpose, they make use of various spices also, by which taste level of their food gets increased to considerable extent and their appetite level also gets increased. However, some people like to consume some vegetables in their raw form.

Human beings can sustain their life because they consume food items of different kinds. If human beings do not get food for a long period of time, then they can die. This is equally true for all the living organisms existing on the earth. It is because of this reason that food and role played by it has remained an important subject of study since a long period of time. It has been by the experts that most of human beings do not eat to sustain their physical survival, instead they eat food as their experience feeling of appetite, which in simple terms is the desire to eat. Human beings or all the living organisms consume food items when some inner signal tells them that it is time to eat. It is possible that a person can feel appetite even after consuming the food.

Varied kind of differences are found in the liking and disliking of people regarding food items. Not only this, different human beings take food in varying quantities. In reality, such habits or features are influenced to considerable extent by various factors. Feeling of appetite can get developed in human being when he is not hungry, which can take place as a result of sensory stimulation such as smelling, seeing and tasting delicious food items. Large number of people can be found who remain in a constant struggle to find the right balanced position in between eating to maintain body function and eating to satisfy their appetite. The factors which play important role in defining what a person is nutritionally are personal preferences, habit, availability of food items, emotional comfort, values, kind of body structure and nutritional values.

Thus from above discussion it can be said that there exist differences in the concepts of hunger and appetite. Hunger is the feeling which is associated with physiological needs of the human beings, while appetite is the desire to eat, which normally accompanies hunger and is psychological in nature.

Human beings consume various kinds of food items in order to satisfy their appetite. Although main function of food is to satisfy the appetite, but various other kinds of functions are also performed by it, some of which are as follows:—

a. It is only through the energy provided by food that human body can perform various kinds of mechanical functions.

b. Human body is composed of various muscles and cells. While performing physical works, some of the cells existing in body gets destroyed, for repair and development of new cells, energy is needed. Such energy for this purpose is provided through the food consumed by human beings.

c. Human body is a highly complex machine, in which various kinds of functions keep on taking place. Various kinds of chemical reactions take place in human body, by while body temperature gets increased. The energy which is required to keep body temperature under control is provided by the food items consumed by human beings.

Thus, apart from satisfying the appetite, food consumption performs various important kinds of functions, for which, human beings should consume proper kind of food in adequate quantity and at proper time. It is important to mention here that generally people believe that if they will consume food at short period of intervals, then energy level will get increased to considerable extent in their body, which is far from true. With such practice, extra energy will get stored in the form of fats under the skin of body, by which body will become over-weighted and person will get affected by the disease known as obesity, which in turn leads to various kinds of problems related to cardio-vascular system of the body.

Experts have provided some basis on which food

items consumed by human beings can be divided into various categories. The basis used for this purpose are : on basis of source, on basis of function and on basis of chemical nature. On basis of source of the food, food items can be divided into two categories, namely, vegetable foods and animal foods. In the category of vegetable foods, items such as rice, wheat and similar kinds are included which are generally sown in the farms, while in second category, those food items are included which are provided by nature, e.g., meat, egg and fish. Likewise, on the basis of functions performed by food items, they can be divided into three categories, namely, body builders, energy yielders and protectives. Those food items which perform the function of body building are included in the first category, in which items like meat, milk and fish are included, while items through which human body gets energy to varying degrees are included in the second category, which generally include the items like cereals and sugars, while food items which help in protecting human bodies from various diseases are included in the third category, in which items like vegetables and fruits are included.

After discussing these concepts at length, now it is reasonable to throw some light on the next concept under discussion, that is nutrition.

Concept of Nutrition

Science of food and it's relationship to health is known as the nutrition. In other words, it can be said that nutrition is the science that investigates the relationship existing in between the physiological function and the essential elements found in the food items consumed by living organisms. The constituents of food that sustain living organisms physiologically are called nutrients, and in this category proteins, carbohydrates, fats, vitamins, minerals and water are generally included. It is necessary for living organisms, and specially to human beings to take such food in which all the nutrients exist by which proper level of all of them can be maintained in the human body. In lack of any of the nutrient in the body, human beings can get attached by various diseases, for which it is necessary to make those provisions by which their body can get all the essential nutrients in proper quantity. Such kind of food which consist of all the essential nutrients in adequate quantity is called Balanced Food.

Unit of measure that indicates the amount of energy one obtains from a particular food is known as calorie. It is in the forms of proteins, fats and carbohydrates that calories are eaten by human beings. Other remaining nutrients are considered to be necessary for proper body function, but no calories to the diet is being added by them in any way. Situation of overweight gets developed as a result of excess calorie consumption. It is no so much the quantity of food human beings eat that is likely to cause weight problems and resultant diseases as it is the relative proportion of nutrients in their diets and lack of physical activity. It has been found from various studies that food which is generally consumed by majority of people in our nation contains high quantities of fat and carbohydrate and as people do not have enough time to get participated in the physical activities as a result of which their body weight gets increased to considerable extent, because of which they become pray of problem known as obesity, which lead to development of various other problems. Experts are of the view that most of the diet related diseases have resulted from the excess calories and increased consumption of fat by the human beings.

Essential Nutrients

As said earlier that it is very important to have some of the essential nutrients in desirable quantities for proper development of the body. As these nutrients are of such importance, thus, it is necessary to discuss them at length, mention of which are as follows :—

Proteins

This nutrient is the most abundance substance in the human body. They are considered to be complex nitrogenous compounds. In reality, proteins are major components of nearly all the cells present in human body. Experts consider proteins to be body builders as they play important role in developing and repairing bone, muscle skin and blood cells. Here it is important to mention that human body is made up of various muscles and cells, which gets damaged while performing different physical functions. For proper functioning of the body, it is necessary to have these damaged cells and muscles repaired, and this important function is performed by proteins found in the food, as a result of which they are considered to be of utmost importance. Not only these, proteins are also key elements of the antibodies which help the human body to be protected from various diseases. Chemical activities taking place in the human body gets controlled by proteins and they also control the hormones which play important role in regulating various functions performed by human body.

As known that various kinds of nutritious elements are obtained through food consumed by human beings, which is required by all the body parts. The function to transport all these nutritions to various parts of the body is also performed by the proteins. Proteins also play important role of supplying other sources of energy to cells situated in different parts of the body in situation when fats and carbohydrates are not available readily. Thus it gets proved by the discussion that adequate amounts of protein in the diet are vital to many body functions and ultimately to survival of human beings.

Proteins consumed by human beings are being broken into smaller molecules by the body parts, which are called amino acids. These parts are linked together like beads in a necklace, and there are total twenty in numbers, out of which three combinations are considered to be most essential for the body. Dietary proteins by which body gets all of the essential amino acids is termed as complete protein. Generally proteins obtained through animals products are considered to be complete in nature. Whenever human being consume food items in which some of the essential amino acids are found in less quantity that desired than whole amount of protein that can be synthesized from other amino acids gets reduced to considerable extent. Proteins that present in the digestible form are considered to be complete in nature.

Proteins obtained through plan sources are found to be incomplete in nature, as they generally lack one or two of the essential amino acids. Experts have divided plant sources of protein into three categories, namely, legumes, grains and nuts. In the first category, food items like beans and corn are included while in the second category falls various nuts. Likewise, various seeds are included in the third category. Certain vegetables also contribute valuable plant proteins. Human beings can get essential amino acids necessary to ensure adequate protein absorption through mixing two or more food items with each other. Incomplete plant proteins with complete low fat animal proteins can be combined by those people who are interested in obtaining all of their protein from plants. One can obtain high quality proteins through low fat cheese, skim milk and egg whites.

Although human beings should eat enough protein, but it should be ensured that you do not consume too much, as it can place added stress on the liver and kidneys. By it, calcium excretion in urine can also be increased, by which risks of bone fractures can get elevated. As cases of persons suffering from various diseases arising as a result of high amount of carbohydrates have come into notice, which has led to production of various food items which have low calories by which high levels of carbohydrates can be eradicated and people can be induced to consume food items rich in proteins. Although diets that deviate from a balanced nutritional approach are almost certainly flawed, most would not cause major health threats in otherwise healthy people when used for short periods of time. High protein diets should be avoided by those persons who suffer from kidney or liver problems.

If a person is fighting off a serious infection or is recovering from surgery or blood loss, then high amount of protein will be required by him. In such kind of situations, proteins that are lost to cellular repair need to be replaced, which can be done by proteins effectively. Protein deficiency continues to pose a threat to the global population, and large number of people in our nation also suffer from this same problem. Experts have recommended protein intake for the average man is only 63 grams, and the average woman need only 50 grams. But still number of people suffering from low level of proteins in their food is increasing with greater pace.

Vitamins

These are considered to be one of the most potent

and essential organic compounds for human body. These nutrients are required by human body in very small amount. No energy is being provided by this nutrient, but it helps in making proper utilisation of other nutrients found in the food. It is necessary to provide vitamins to the body through food as they cannot be synthesized by the body themselves. Vitamins help in maintaining nerves and skin every minute and every day. Human body requires blood cells without which various body parts will not be able to function properly. Vitamins help in making the development of blood cells in the body possible.

Human beings meet various kinds of accidents, as a result of which different kinds of wounds get developed on their body. In the absence of adequate amount of vitamin in the body, it will become quite difficult to heal the wounds. Vitamins found in the body can be destroyed by various factors, some of which are age, heat and various environmental conditions.

Categorization of Vitamins

Experts have classified vitamins obtained through various food items into two categories, namely, fat soluble and water soluble. Fat soluble vitamins get absorbed through intestinal tract with the help of fats, while water soluble vitamins get easily dissolved in water. Vitamins A, D, E, and K are grouped under the first category while Vitamin B and Vitamin C are placed under second category. Fat soluble vitamins are tend to be stored in the body and toxic accumulations in the liver may cause cirrhosis like symptoms which water soluble vitamins are generally excreted and cause few toxicity problems in the body.

Concept of vitamins cannot be understood properly without mentioning the sources through which various vitamins can be obtained and the important functions performed by them in length, mention of which are as follows :—

- Vitamin A : This vitamin is an unsaturated alcohol in reality. Carotene is the precursor of this kind of vitamin, which exists in three forms, namely, alpha, beta and gamma. This vitamin is considered to be very important for human body, as it performs various kinds of functions, some of which are as follows :—

i. It helps in production of retinal pigments in the body, with the help of which human beings become able to view things even in very dim light.

ii. It helps in maintaining the integrity of epithelial cells found in the human body.

iii. It saves human body from getting affected by various infections.

iv. It helps in getting the skeletal system of the body developed.

Importance of this kind of vitamin can be analysed from the fact that if found in deficient amount, then human body can get affected by various serious kinds of diseases some of which include Night Blindness, Bitot's Spots and Keratosis. It is possible to get this kind of vitamin from various food items, some of which are carrot, spinach, green vegetables and mangoes. It can be obtained through consumption of various diary products like milk, butter and cheese.

- Vitamin B Complex : Water soluble vitamins belonging to group of Vitamin B complex include the following :

i. Thaimine: It is also known as Vitamin B_1. One can obtain this kind of vitamin from food items like peas, beans, oatmeal, vegetables and fruits. A person having deficiency of this vitamin can get affected by the diseases known as Beriberi and Wernicke's Encephalopathy. An individual requires 2 mg of this kind of vitamin daily, which gets increased for the persons who take carbohydrate in more quantity.

ii. Riboflavine: It is also known as Vitamin B_2. One can obtain this kind of vitamin through milk, eggs, liver, kidney and green leafy vegetables. Importance of this kind of vitamin can be analysed by the fact that deficiency of this vitamin can lead to problem of Angular Stomatitis. Not only this, person having this vitamin in deficient quantity will be more susceptible to cataract. A healthy human being requires 2 to 3 mg. riboflavine in the food daily.

iii. Nicotinic Acid: Also known as Vitamin B_3. The main sources through which this kind of vitamin can be obtained include the meat, poultry fish and ground nut. A person having this vitamin in deficient amount can get affected by problem known as Pellagra. Generally it is found that problem of Pellagra gets developed more in the persons who consume maize or jowar in higher quantities. Person suffering from such disease should take a mixed diet which should consist of milk and meat while maize and jowar should be avoided by such person to maximum possible extent.

iv. Pyridoxine: It is also known as Vitamin B_6 is often found in the food items like milk, egg yolk, fish and vegetables. An adult requires two mg. per day of this vitamin and it's requirement gets increased for a pregnant woman, during which it rises upto 2.5 mg per day. Deficiency of this vitamin can lead to various problems in human beings, some of the important out of which are irritability, loss of body weight and shortage of red blood cells in the body. Adults having this vitamin in deficient quantity will face the problems of lesions in their skin and mouth. One of the most important and serious kind of problem arising out of deficiency of this vitamin is Peripheral Neuritis, for avoidance of which, it is very important to take such food which consists of this kind of vitamin in proper quantities.

v. Cyanocobalamin : It is also known as Vitamin B_{12} and the main source from which it can be obtained are non-vegetarian food items, like liver, kidney, eggs and milk. This kind of vitamin is also present in vegetable items, however, in very less quantity. The daily requirements of this vitamin is 1 to 1.5 mg. per day and it's deficiency can lead to various serious consequences. For instance, a person having this vitamin in deficient or less than required quantity can get affected by the disease known as Megaloblastic Anemia. Thus, this vitamin plays an important role in producing red blood cells in the body.

vi. Folic Acid : This vitamin is also very important and essential for human beings as it's deficiency can lead to defective maturation of red blood cells in the body, by which problem of megaloblastic anemia can developed in the human body. The main features of Megaloblastic Anemia is that large sized blood cells begin to be released in the body, which are known as megalo blasts, however, there is less amount of hemoglobin in such blood cells. A health adult requires 100 to 300 mg. daily in the food and this can be obtained through vegetarian foods, whereas it exist in non-vegetarian foods in less quantity.

- Vitamin C : This vitamin is also known as Ascorbic Acid. One can obtain this vitamin from citrus fruits, tomatoes and green vegetables. One of the main richest source of this vitamin is guava fruits. All the human beings require this vitamin in amount of 40 to 60 mg daily. This vitamin is very important for human beings, which can be proved by the fact that it's deficiency can lead to development of disease known as Scurvy, by which muscles and joints of the body become very weak and person feel fatigue within short period of time. Not only this, a person suffering from this problem will have bleeded gums and weak teeth.

- Vitamin D : This kind of vitamin occurs in two forms, namely, Calciferol and Cholecalciferol. Generally Calciferol is known as Vitamin D_2 while term used to denote Cholecalciferol is Vitamin D_3. Vitamin D_2 can be obtained through irradiation of ergosterol, while Vitamin D_3 is found in animal fats and fish liver oils. This kind of vitamin can also be obtained through exposure of cholesterol of skin to sunlight, which consist of UV rays, by which important role is played in production of this vitamin. The importance of this vitamin can be analysed from the fact that a person having it in less than desired quantity can suffer from the problem known as Rickets. Disease known as Osteomalacia also gets developed as a result of deficiency of this kind of vitamin in the body.

Vitamin D performs various kinds of important functions, mention of some of which are as follows :—

a. Absorption of calcium and phosphorus begin to take place at increased pace in the intestine through this vitamin.

b. Human kidneys become able to absorb more amount of phosphate with the help of this vitamin.

c. It plays important role in the formation of bones and teeth in the body.

d. Without it, human body parts cannot grow at normal pace.

Vitamin E : This kind of vitamin is part of group of compounds, which is known as tocopherols, which are generally divided into categories of alpha, beta, gamma and delta tocopherols. Among them, alha tocopherol are considered to be most active in nature. A human being requires this vitamin in the quantities of fifteen to thirty mg daily. The importance of this kind of vitamin can be analysed from the fact that deficiency of it can produce problems of abortion and sterility in animals, however, no such symptoms have been established in human beings till now. One can obtain this vitamin through soya bean oil, wheat germ oil and rice germ oil.

Various kinds of important functions are performed by this vitamin in the human body, mention of some of which are as follows :—

a. It is only through this vitamin that animals and living creatures can perform the task of reproduction.

b. In the muscle metabolism, this vitamin plays very important role. Through this vitamin, it becomes possible for human muscles to store creatine in adequate quantities.

c. Such vitamins consist of antioxidant effects, as a result of which they play an important role in saving Vitamin A from getting destroyed.

Although all the functions mentioned above are of equal importance, but function of antifertility effect is considered to be one of the most important of them for human beings.

Vitamin K : This vitamin is found in two forms, namely, Vitamin K_1 and Vitamin K_2. This is also one of the important kind of vitamin, which can be realised by the fact that it's deficiency can lead to defective blood clotting mechanism, as a result of which, if person gets any kind of wound, then lot of blood will flow from his body. The recommended value which is to be consumed everyday by human beings should be 0.03 mg per kg. The main sources through which this kind of vitamin can be obtained include green leafy vegetables, cow's milk and spinach. Intestinal bacteria play important role in synthesising Vitamin K_2.

Carbohydrates

The energy by which human beings become able to sustain various activities is obtained through nutrient known as carbohydrate. In comparison to other nutrients, carbohydrates can be metabolized more quickly and efficiently. Human body gets energy from carbohydrates immediately, the main reason of which is that it gets converted into glucose very easily. Through this process, cells existing in human body gets the fuel required for their proper functioning. Food rich in carbohydrate play an important role in the functioning of internal organs, specially the nervous system and muscles situated in the human body.

As carbohydrate provide both an immediate and a time-released energy source because of which they are considered to be best fuel for endurance athletics. They get digested easily and then get metabolised consistently in the bloodstream. Experts have divided the carbohydrate into two categories, namely, simple sugars and complex carbohydrates. Carbohydrate in the form of simple sugar is found primarily in fruits, the most common form of which is glucose. Complex carbohydrates are found in the grains, cereals and yellow fruits apart from various vegetables. Human cells get energy through the function in which all kinds of simple sugars get converted into glucose.

As told that most common form of simple sugar is glucose. This glucose is sweet in taste and can be obtained through various substances like honey, vegetables and fruits. Another form in which simple sugar exist is Fructose, which is obtained through fruits and berries. Both of these forms of simple sugars are monosaccharides in nature, which imply that both of them contain only one molecule of sugar.

When two monosaccharides get unite with each other, then the concept of Disaccharides get developed. The most immediate and simple example of this kind of food item is table sugar which consists of a molecule of fructose which is bounded chemically to a molecule of glucose. Another important form of disaccharide is lactose, which is found in milk and various milk products. Lactose is formed by the combination of glucose and galactose. Prior to their use by the body, disaccharides are being broken down into simpler units by the human body.

Long chains of saccharides form another important kind of complex carbohydrate, which is known as Polysaccharides. Before utilisation, they are also broken down into simple sugars by the body. Majority of the complex carbohydrate group is formed by the starches, which are obtained by human body through flours, breads and potatoes. Such matter gets stored in human body in the form of fat, and this function is performed by muscles and liver in a polysaccharide form called glycogen. Whenever need of high energy is felt by the body, glycogen are broken down in the form of glucose immediately.

Fibre is one of major form of complex carbohydrates and it is the ingestible portion of plant foods that helps move foods through the digestive system and softens stools by absorbing water. These also helps in controlling body weight as feeling of fullness of stomach is developed by them. Experts have divided fibres into two categories, namely insoluble fibres and soluble fibres. Soluble fibres appears to be an important factor in lowering blood cholesterol levels, by which risk for various cardiovascular diseases get lowers down, while insoluble fibre is associated with various gastro-intestinal benefits and play an important role in lowering down the risks for several forms of cancers. One who wants to increase his intake of dietary fibre should eat more complex carbohydrates, however, this should not be done without due planning, as sudden increase in dietary fibre can cause various kinds of problems or difficulties, some of which are cramping and intestinal gas. Such side effects can be reduced to considerable extent by consuming plenty of water or other liquids in higher degree. Although fibres should not exist in higher quantities in the body, but still various good effects have been come into notice by the studies conducted by experts. For instance, it has not proven that fibres help in providing protection against some specific kinds of cancers. Not only this, it also provides protection against problems like constipation, heart diseases, diabetes and obesity. For increasing fibre intake, one should eat a variety of food, should consume less processed food, eat skins of fruits and vegetables, should spread out his fibre intake and get the fibre from various food items and not from medicinal pills.

Fats

These are also known as lipids and forms one of the basic nutrients required by human body for proper development and growth. Fat plays an important role in keeping the skin and hair healthy. It insulate various body organs against shock. An important role is being played by it in keeping the body temperature under control and in promoting healthy cell function. Through fats, taste of food items become better and it is through this constituent that food attains the quality of deliciousness. If there is absent of carbohydrate in the sufficient amount in the body, then concentrated form of energy can be provided by this nutrient.

Minerals

These are the inorganic and indestructible elements that provide help and assistance for various physiological processes taking place within the body. Human body consists of more than fifty chemical elements, which are being obtained through various food items. Around 1/20th part of body weight is made up by the minerals. It is not possible for body parts to get vitamins absorbed in the absence of minerals. Experts have divided minerals found in human body into various categories, two of which are Macrominerals and Trace Minerals. Macrominerals are those minerals which are required by human body in large amounts, for instance, sodium, calcium, phosphorus and potassium, while in the Trace minerals includes those elements which are required in trace amount by the body. Zinc, iron and copper are some of the examples of this kind of minerals. Although it cannot be denied in any way that minerals are necessary for proper body functions, however, they should be consumed within restricted quantities.

Sodium

This mineral is necessary for regulation of blood and body fluids. Functions of transmission of nerve impulses and heart activity are controlled by this mineral. This help in various metabolic functions taking place in the human body. It has been found that majority of people consume sodium in the quantity which is much more than the needed. Experts are of the view that an adult who does not sweat profusely needs very little amount of sodium, which has been set out to be 500 mg. per day.

One of the most common and important source of obtaining this form of mineral is table salt, however, table salt accounts to very little per cent of sodium intake, while the remainder comes from water and from highly processed food items which are infused with sodium and help in enhancing the flavour of food. Hundred milligrams of sodium per serving are found in the food items like pickles, salty snack foods and processed cheese. From various studies it has been found that there is a positive link between the excessive sodium intake and high blood pressure. In other words, it has been found that persons who take more salt in their diet generally face the problem of high blood pressure. To reduce risk for cardiovascular disorders, it is necessary to cut back on the sodium. It has also been prove by various studies that calcium loss in urine can take place as a result of high sodium intake, by which risk for debilitating fractures get increased with increase in the age.

Iron

It is considered to be one of the most important nutrient required by human body. Around three to four grams of iron is found to be present in body of an adult, out of which, a large proportion of which exists in hemoglobin present in the blood. There is around 3 to 3.4 mg of iron in each gram of hemoglobin. Iron deficiency is found to be one of the most common kind of nutrient deficiency, by which majority of people not only in our nation, but different parts of the world are being affected. This problem is more alarming in the developing countries like ours, majority of sufferers of this problem are children and women. When human body becomes unable to produce hemoglobin, then a disease called anemia gets developed in the body. In people affecting with iron deficiency anemia, a condition known as pica can also get developed which is an appetite for ice, clay and other non-food substances that do not contain iron and in fact may inhibit iron absorption to considerable extent.

In the person having deficiency of iron, various kinds of problems are faced. Generally it is found that body cells present in such person's body receive less oxygen as a result of which, carbon dioxide wastes are removed from body less efficiently. As a result of this condition, a lot of tiredness is felt by such person. Generally women are more likely to develop iron deficiency problems as they typically eat less food than men and their diets consist of less iron elements.

Significant role played by mineral in the form of iron can be analysed from the fact that it's deficiency can tax the immune system, causing it to function less effectively. Iron deficiency can also lead to immunity and decreased resistance to various infections. There appears to be a slight association between iron deficiency and cancer.

Calcium

This nutrient plays very important role in human body, some of the functions performed by it are as follows:—

a. It helps in building strong bones and teeth.

b. It helps in regulating the function of muscle contraction.

c. It provide assistance for various cardiac actions.

d. In the retina, light is being transformed into electrical impulses through this mineral.

e. It helps in keeping control over the heart beats.

f. It helps in keeping the fluid balance within the cells existing in human body.

Over half of the calcium intake usually comes from milk, which is one of the highest sources of dietary calcium. Other good sources of calcium are green and leafy vegetables. Cauliflower and many peas also offer good supplies of calcium to the body. Good amounts of calcium exists in various nuts, particularly almonds. Some fruits, such as citrus fruits have moderate amounts of this mineral.

One should not consume carbonated soft drinks in larger quantities as these consist of added phosphoric acid which can cause you to excrete extra calcium which may result in calcium being pulled out of your bones. Chances of development of kidney stones gets increased with existence of calcium quantities in much larger quantities than required.

Fluorine

This kind of mineral is considered to be one of the abundant element found in nature. This element in highly reactive in nature, as a result of which it occurs only in a combined form in form of fluoride. Major portion of fluoride in the body is found in the bones and teeth. Experts have recommended the intake of 0.5 to 0.8 mg per litre of fluorine in the water for a healthy adult. Various kinds of dental problems get arise as a result of deficiency of fluoride in the body, however, if taken in excess than required quantity, then it can lead to dental and skeletal fluorosis in the body.

The main sources by which fluorine can be obtained are drinking water, sea fish, cheese and tea. This mineral should be taken in the right or adequate amount as various kinds of important functions are being played by it. For instance, it helps normal mineralisation of bone and also helps in formation of dental enamel.

Iodine

This is considered to be one of the most important micro-nutrient. Generally it is present in the human body in quantity of 50 mg. Sea foods of various kinds are considered to be important source of iodine. In meat, cereals and vegetables also this mineral is found in moderate quantities. The best source of availability of this mineral is consumption of iodised table salt.

This mineral performs various kinds of important functions in the body. For instance, it helps in the synthesis of thyroid hormones in the body and it is also required for proper development of various body parts. It is very necessary to consume this mineral in adequate amount as it's deficiency can lead to various kinds of problems, one of an important and of serious nature of which is goiter. Not only this, this mineral should be consumed in higher quantities by a pregnant woman as it's deficiency can lead to abortion and mis-carriage.

Copper

This kind of mineral is required for the synthesis of hemoglobin present in the blood. It should be taken in adequate quantity as it's deficiency can lead to defective formation of elastic tissues and tissue matrix in the body. The main sources from which it can be obtained are common salt, tomatoes and green leafy vegetables.

Zinc

This kind of mineral is found in all tissues in form

of traces. It is one of an important constituents of several enzymes. The daily requirement of a human being is five to ten milligram. The sources through which it can be obtained include meat, fish and milk. It is very important to take this mineral in proper quantity as it's deficiency can lead to various kinds of sexual infantilism and skin disorders. Not only this, person having inadequate amount of zinc can suffer from problem of slow growth process as a result of which, he can become physically or mentally regarded.

Water

Human beings can survive for much longer without food than they can without water. Even in severe conditions, the average person can go for weeks without certain vitamins and minerals before experiencing serious deficiency symptoms. However, dehydration can cause serious problems within a matter of hours, after a few days without water, death is likely. Here it is important to mention that the condition in which water level gets lower than the adequate or required in the body, it is called dehydration.

Various studies show the fact that a health human being requires around six to eight glasses of water per day. Because of high concentrations of water in most of the food items human beings consumed, however, the actual number of glasses needed each day becomes somewhat less than this. On various factors individual need of water of human beings depend, some of which are age, body size and environmental temperature. Higher volume of fluid intake becomes necessary for persons suffering from certain specific diseases like diabetes. However, it should be born in mind that if clean water can prove be nectar for the human beings, dirty water can play role of poison for them. It is very necessary to take only clean drinking water, otherwise, various kinds of diseases will get arise in the body. Not only this, there are various infections which result as a result of drinking impure water. All the provisions should be made to keep the drinking water safe and clean. Firstly clean utensils should be used to store such kind of water and proper attention should be paid to the fact that no one puts his fingers in such water. Number of persons suffering from problems arising as a result of drinking dirty and unclean water are increased in our nation, the main reason of which is that no proper provision of safe and drinking water has been made in various parts of the nation inspite of various steps taken by government authorities in this direction. But still, significance of water as one of an important nutrient of the body cannot be ignored in any way. This should be understood by the politicians and service providers that without providing clean drinking water to the citizens, they even cannot hope to raise the level of their health in any way.